THE *Planetary*
MIND

A R N E   A .   W Y L L E R

THE *Planetary* MIND

MACMURRAY & BECK
ASPEN, COLORADO

Library of Congress Cataloging-in-Publication Data
Wyller, Arne A.
    The planetary mind / Arne A. Wyller.
        p.   cm.
    Includes bibliographical references and index.
    ISBN 1-878448-64-1
    1. Evolution. 2. Philosophy of mind. 3. Philosophy of nature.
I. Title.
B818.W95   1996
213—dc20   95-18014
                                                                  CIP

The Planetary Mind designed by Susan Wasinger.
The text was set in Berkeley by Pro Production.
Project management by D&D Editorial Services.

TO ANNE-MARIE, *who gave me life,*
AND TO VERONICA, *who gave me life anew*

C O N T E N T S

ILLUSTRATIONS

ACKNOWLEDGMENTS

So often acknowledgments involve only direct, personal contacts with living human beings. I should like to start out, however, by acknowledging my immense indebtedness to the individual humanistic, scientific, and religious voices, past and present, who have left their thoughts and articulations in the numerous works cited in the bibliography of this book. Without their musings, ponderings, and statements of observational facts, my personal weave of thought would be for naught.

However, a scientific-philosophical work like this—written with a lay audience in mind—requires patient midwifing by several persons. The first version was written out in longhand during spring 1987 in Santa Fe and patiently deciphered and typed by my daughter, Unni Wyller, that same year. That scientifically oriented version lingered rejected for several years.

Then in 1992 I happened to find my first editor, Stephen Shapiro, who painstakingly simplified my earlier manuscript. Tanya Wells excellently retyped the manuscript on her trusty Macintosh® and by a circuitous route this new version came during the winter of 1993–1994 into the hands of the present editor, Fred Ramey of MacMurray & Beck. This event brought about a dramatic revision and expansion of the entire manuscript into its present form. For this I cannot thank Fred Ramey enough for his inspiring criticisms and suggestions, which sparked many new ideas and formulations in my rewriting. Again Tanya Wells, in addition to impeccably retyping countless modifications, also provided expert advice on the art of popular communication. I am also indebted for the privilege of doing much of my expanded studies in the Meems Library of St. John's College in Santa Fe.

Further enhancement of the clarity and consistency of the manuscript came through the diligent copy editing of Phil Murray and the artistic work of designer Susan Wasinger and technical illustrator Susan Strawn, together with the project leadership of Deborah Lynes and Darice Whetstone. It has been a truly rewarding and exhilarating experience to work with this team.

To all of the above I wish to express my deep gratitude for aiding in the completion of this soon-to-be ten-year quest of mine.

Arne A. Wyller

SANTA FE, NEW MEXICO

OCTOBER 1995

PROLOGUE

*T*ime magazine's year-end issue for 1992 had a cover story written by Robert Wright, senior editor at the *New Republic,* with the title "Science, God and Man." At the beginning of this essay Wright makes a distinction between conventional and unconventional religiousness. Unconventionally religious people, according to Wright, are "religiously inclined, but reaching for scientific support. . . . They kind of believe in some deity, but would not mind seeing some hard evidence."

This book addresses this audience of the religiously unconventional. I believe that hard evidence for a kind of deity appears in the creation of the *information content* in all life forms, past and present, on this planet Earth. However, I wish to make it perfectly clear that my book has nothing to do with creationism or conventional New Age thinking. As a scientist, I seek to examine and marvel at the amount of sophisticated technical information deposited in our planetary life forms. The central point in my argument is that there has not been time enough for this information to have been deposited into the genetic blueprints by chance.

I believe that we shall see a new paradigm emerge in the biological sciences that will incorporate the idea of a *Planetary Mind Field,* adjunct to and coexisting with the energy fields of the physicist that create matter. This Mind Field obeys the physical laws of the material world, yet is able to program genetic blueprints into all biological species.

This new paradigm will include some of the tenets of Darwinism and neo-Darwinism. It will unconditionally accept the concept of the evolution of life over geologically established time. It will incorporate the neo-Darwinistic

concept of populations and natural selection of especially well-adapted individuals of a given species. But it will abandon the Darwinian notion of
chance and natural selection as the mechanism for creating *entirely new
species*. That, in the new paradigm, is the prerogative of the Planetary Mind
Field.

My purpose is to open various "windows" of current knowledge in biology and physics to the lay reader and to make the reader aware that growing
numbers of physicists and other scientists are deeply dissatisfied with the
current intellectual atmosphere in which science persistently refuses to consider the issues of mind and consciousness. I have tried to structure the
book in the spirit of books like Carl Sagan's *Cosmos* (without any mathematical explication).

I hope the book will be of guidance in the reader's personal quest for
meaning. We are approaching the year A.D. 2000. Are our seemingly disparate windows of knowledge going to be put together into a new worldview
for our age? It is only ten generations since Western civilization gave birth to
modern science, and it is only one hundred generations since written civilization began in ancient Sumer. Today finds the modern Western world in
an age of confusion, bewilderment, and search for meaning. To many of us,
ours is an age of fear—fear of nuclear war, fear of pollution, fear of humanity's physical and moral self-destruction. Our age is similar in many ways to
the late phases of the decline of the Roman Empire. There, in the fourth century A.D., paganism took many forms (Mithraism, Neo-Platonism, Stoicism,
Cynicism, and the local cults of municipal and rustic gods). Spiritual an-
guish and bewilderment were prevalent, both vividly attested to by the he-
donist, man-of-the-world, later Saint Augustine (353–430) in his many written works.

Augustine found a way out of his personal dilemma by embracing the
religion of Christianity. To many today, no one religion appears a viable alternative in light of the social issues of our planetary family and the findings of modern science. As a result, many of us live in a religious-philo-
sophical vacuum, filled with uncertainty, despair, and gloomy forebodings of
global doom under the so-called nuclear umbrella.

To a small minority of concerned scientists and humanists, however, it appears that we are witnessing not so much the demise of humanity as the coalescence of a human global consciousness on all levels. This is a consciousness groping and striving for a coherent new worldview in harmony with the exhilarating new findings in modern science, be they in the hard sciences like physics or soft sciences like biology. Ours is an age akin to the birth of the Western Renaissance around A.D. 1200, a time when individuals tried to formulate new conceptual structures based on observational experience rather than on supernatural dogmas. The modern counterparts of those individuals are searching today for a new set of basic assumptions on which to explain reality.

In that sense, our age shares the agonies and the exhilarations of the early scientists and philosophers of the Renaissance, who struggled almost obsessively with the formulations and creative use of the concepts of motion, acceleration, impetus, and force. Those basic concepts and their mathematical definitions required an immense intellectual effort both to be formulated adequately and to be accepted. But today they are part and parcel of our everyday language, just as are the concepts of electricity and magnetism, which arose out of the eighteenth century through the efforts of people like Benjamin Franklin.

Some of us believe that current struggles for a new paradigm will inevitably have to involve the life sciences, as well as new concepts for explaining the *evolution* of the life forms on our planet. We shall have to painfully but exhilaratingly make repeated and renewed attempts at putting evolution in a richer conceptual structure than is now the accepted canon. This book is an attempt at such lateral thinking, although I am fully aware of the many pitfalls in such an undertaking and the criticism that will be leveled at it.

Yet, as I will endeavor to show, such attempts have been made repeatedly throughout intellectual history from Aristotle and Plotinus through to Spinoza, Bergson, Whitehead, and Teilhard de Chardin. Just as the idea of the atom and the ideas of alchemy needed over 2,000 years to bear useful fruit in modern chemistry and nuclear physics, so the idea of the amalgamation of spirit and matter does perhaps now, after centuries of debate, approach a stage of acceptance in human intellectual pursuits.

It has often been observed that in our century, possibly during the last generation, there have been more scientists than all scientists taken together in history. Certainly, the twentieth century has witnessed an explosion of both theoretical and observational knowledge without parallel in human history. This is the century in which the New Story of Science has been written and is still being written, as surveyed admirably in a book by Robert Augros and George Stanciu (1984).

I believe that the time has therefore come to peer through selected windows of current knowledge in evolutionary biology to try to extract material for the construction of a new paradigm that will involve both spirit and matter. We must attempt yet another synthesis of certain elements of the world of the humanist and the world of the scientist. Although my present attempt may fail, others will try. It is all a part of the patient sculpturing process of cultural evolution, chiseling away excess intellectual waste and erroneous structures until a richer, more variegated image of reality appears.

An element of naiveté may be necessary in the midwifing of such new concepts, especially those that concern the biological sciences and their bearing on the ultimate question of the evolution of life. It is a question most of us face sooner or later in our lives, succinctly put by the French impressionist painter Paul Gauguin: What are we? Where do we come from? To where do we go?

For a professional astrophysicist like myself, at times it feels presumptuous to venture into an entirely new field—that of evolutionary molecular biology—and search therein for new paradigms. In my personal quest, however, I have been driven by the urge akin to that of the young child in H. C. Andersen's classic fairy tale "The Emperor's New Clothes." Confronted with the traditional Darwinian conception, my spontaneous reaction is, "but he has no clothes on." There exists no mathematical validation of Darwinian assertions about the creation of new species.

The modern scientist's dilemma in exploring new paradigms outside his or her narrow professional field is penetratingly formulated by the famous Nobel Prize winner in quantum theory, Professor Erwin Schrödinger. In his classic short treatise *What Is Life?* published in 1944, he states:

A scientist is supposed to have a complete and thorough knowledge at first hand, of *some* subject, and therefore is usually expected not to write on any topic of which he is not a master. This is regarded as a matter of *noblesse oblige*. For the present purpose I beg to renounce the *noblesse*, if any, and to be freed of the ensuing obligation. My excuse is as follows.

We have inherited from our forefathers the keen longing for unified, all-embracing knowledge. The very name given to the institutions of highest learning reminds us that from antiquity and throughout many centuries, the *universal* aspect has been the only one given full credit. But the spread, both in width and depth, of the multifarious branches of knowledge during the last hundred odd years has confronted us with a queer dilemma. We feel clearly that we are only now beginning to acquire reliable material for welding together the sum total of all that is known into a whole; but, on the other hand, it has become next to impossible for a single mind fully to command more than a small specialized portion of it.

I can see no other escape from this dilemma (lest our true aim be lost for ever) than that some of us should embark on a synthesis of facts and theories, albeit with a second-hand and incomplete knowledge of some of them—at the risk of making fools of ourselves. . . .

So much for my apology. [Italics in this sentence added]

Likewise, the approach in this book will be a re-entry, if possible, into an almost childlike state of naiveté. It is an attempt at lateral thinking, off the beaten path, wedding old humanistic concepts into new molds furnished by the modern insights of science. Accordingly, we will examine current scientific thinking in fields ranging from the creation of the Universe through the creation of the Earth, followed by an examination of evolutionary biological "breakthroughs" such as the creation of the cell, the genetic code, protein synthesis, the eye, and the brain. This will be followed by a critical examination of the evolutionary paradigms of Darwinism and neo-Darwinism, including the discipline of self-organization.

In looking through these windows of knowledge, it is my hope to convince you, the reader, of the remarkable—yes, astounding—amount of *high-technology information* that has gone into the creation of life.

The question naturally arises then as to how this remarkably sophisticated knowledge has been written into the genetic blueprints of life. The thesis of this book is that there simply has not been time enough, in the mere 600 million years it has taken to create a conservatively estimated 100 million different species, for this knowledge to accumulate by the Darwinian notions of chance and natural selection; nor can the conventional neo-Darwinian wisdom that's evolved since Darwin's time provide a suitable explanation.

Many scientists strongly believe that the Darwinian theory has outlived its usefulness. I am certainly one of them. In the last decade or so, considerable attention has been given to the *Gaia hypothesis* put forward by British physicist J. E. Lovelock and American biologist L. Margulis. Theirs is the modern formulation of the interrelation and interdependence of all life forms on planet Earth as one coherent single organism. It is not a novel thought: the Hellenistic philosopher Plotinus stated this concept of the world as one animated organism in the second century A.D.

In this book I intend to take the modern Gaia hypothesis one quantum leap further and to posit that *humanity and all other life forms in the past and present lie embedded in an invisible Planetary Mind Field that pervades the entire Earth.* The creative activity of this Mind Field is responsible for the appearance and evolution of all forms of life on planet Earth.

It is my firm conviction that this hypothesis of the existence of a Planetary Mind Field will be forced upon us when we seriously face the issue of how to account mathematically for the sophisticated information put into the genetic blueprints of all multicellular life that has existed on this planet within the last 600 million years.

That this Mind Field is assumed to be invisible should not be reason to raise skeptical eyebrows. We accept the existence of invisible gravitational forces that anchor our feet to the ground. We accept the invisible air we draw into our lungs, not to mention the barrage of invisible electromagnetic signals flying around, only to be made visible and audible by our TVs and radios

A further corollary to this thesis is that learning to unravel the information in the genetic blueprints for the various life forms is in reality *learning to read the genetic instructions written by this Planetary Mind.* If these genetic instructions have been created by chance and natural selection, as the Darwinists wish us to believe, the creation mechanism is trivial. But if in the genetic instructions we read "God's own language," the mechanism is awesome!

We spend millions of dollars trying to establish contact with alien civilizations in outer space, while in reality an alien mind may be found right here on Earth, revealing its existence to us through the life forms it has created over millions of years. Tribal wisdom has always maintained this view of the world. So, too, have poets and religious visionaries. Now may be the time when this ancient wisdom will be synthesized in the biological sciences, to lead us out of the intellectual malaise and impasse that have permeated so much of our century.

To some the hypothesis of a Planetary Mind Field may appear far too radical. But in no way does it constitute an endorsement of the views of creationists. To them, the Bible is to be taken more or less literally, and the natural laws of physics are easily suspended. Furthermore, this book does not subscribe to the notion of a static, old-fashioned, omnipotent God, who can arbitrarily suspend the natural laws of physics that govern matter.

The Planetary Mind Field is itself evolving in terms of cosmic awareness and is using humanity as a primitive "tool" to orient itself and connect itself to the rest of the Universe. Over literally billions of years it has created, under the constraints of the natural laws of physics, new life forms out of organic matter. It has done so by manipulating and creating the instructions in the molecular blueprints that make up the genes in the DNA of the various life forms.

For 99.9 percent of the time of life on Earth, this Mind Field has not been concerned with the problems of humankind because it had not yet created us. However, right now, in this exceedingly brief moment in the history of time, its focus of attention is its creative involvement in the evolution of humanity.

Evolution is now going on in all cultures, in economic and political systems, even in *religions*. My personal conviction is that humankind is evolving into a "junior partnership" with the Planetary Mind Field. Thus, we shall have to accept individual responsibility for solving our domestic problems on this planet, under the guidance of the ideas—moral, religious, and otherwise—injected first into our individual consciousness and thereby into our cultures by the Mind Field.

This partnership will require us to "grow up" cosmically—no more demands for "free lunches" or miracles. We must also come to accept that no single religion can claim to be the unique path to connect to the Mind Field. The "cosmic bottom line" is our collective mind's evolution into religious and political pluralism, with mutual tolerance of diverse belief systems as long as they do not oppress individual freedom.

C. S. Lewis, that master of theological fiction and humanistic science fiction, in his last book, *Till We Have Faces,* states, "The Gods will not speak to us face to face until we ourselves have a face." Perhaps the time is approaching for such an encounter.

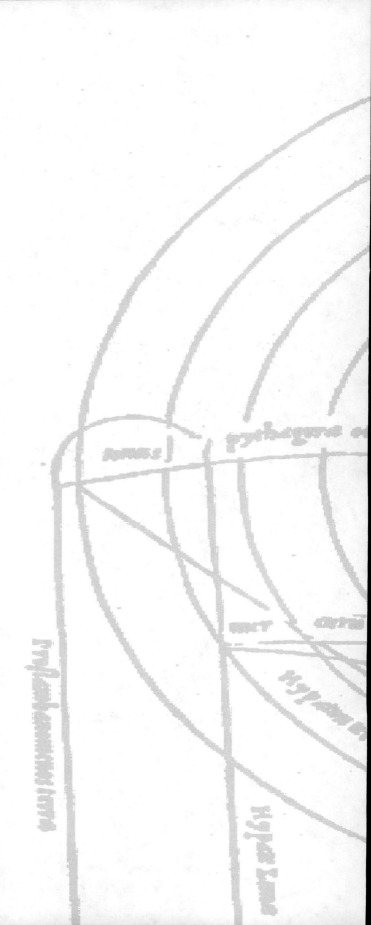

The CREATION *of the* UNIVERSE

The Universe was the size of a grain of sand one hundred-thousandth of a second after the Big Bang began. It was the size of a hazelnut only *one second* later. Those theoretical sizes may be fortuitous—or they may be remarkable interstices between the world of the mystic and that of the natural scientist. The words of the fourteenth-century hermitess Julian of Norwich could serve admirably as a statement on the early stages of the Big Bang model of the Universe: "He showed me a little thing, the quantity of a hazelnut, in the palm of my hand, and it was a round ball. I looked thereupon with eye of my understanding and thought: what may this be? And it was answered generally thus: it is all that is made."

Where does the Universe end? When did the Universe begin? How did it all start? These are some of the questions intelligent human beings have asked throughout the millennia. Many and varied are the answers given in the creation tales of primitive tribes and advanced civilizations.

The modern Western creation myth is no different. In contrast to the myths of earlier ages, this one may survive our times because it is anchored in some undisputed observations of large-scale characteristics of the Universe that were not accessible to previous generations. Still, it is important to examine some of the arguments and evidence for the current tale we tell of the creation of our Universe. The story is less than seventy years old and is fashioned out of the interplay between the astronomer's observations and the physicist's theoretical mind. It is a story awesome in its monumentality and beautiful in its simplicity. Although it has been popularized countless times in the last decades, it bears retelling because it will help us place ourselves in the larger cosmic environment before we begin to look at our Earth and the life it holds.

The twentieth-century creation myth has evolved through the intellectual contributions of people from a wide variety of cultures. Our myth arises out of an emerging chorus of diverse voices—first the theoreticians and then the observers. One voice—that of Albert Einstein—set the theme in 1917 with the publication of his pathbreaking paper "Cosmological Considerations of General Relativity."

The theme of Einstein's story is to consider theoretically the evolution of a finite Universe filled with a smooth distribution of matter. There are no lumps or clusters in Einstein's vision of the Universe, but rather the same density everywhere. In such a simplified model, the natural tendency would be for matter to fall inward toward a common center of gravity under the influence of that center's gravitational pull.

Since Earth has existed for thousands of millions of years, evidently no total collapse of the Universe has occurred during this period. For his theory of the development of the Universe to account for the absence of such a collapse, Einstein felt driven to introduce arbitrarily a *repulsive* force to counterbalance gravitation. He himself stated later that this was the "greatest blunder" he ever committed. Also arbitrarily, he regulated the strength of this repulsive force so that it exactly counterbalanced gravitational pull. Einstein's model of the Universe was static; it neither collapsed nor expanded.

After Einstein's paper all through the 1920s theoretical cosmologists, like Dutch astronomer Willem de Sitter, Russian mathematician Alexander Friedmann, and Belgian abbé Georges Lemaître, worked on refinements of Einstein's mathematical theory of the Universe. Each of these theoreticians contributed vital pieces to the current prevalent view of the creation of the Universe.

De Sitter's great contribution (also in 1917) was that he showed mathematically that Einstein's model for the Universe would become unstable at the slightest provoking disturbance—say, the collapse of matter to form a star. As a result of this instability, the Universe would then expand, such that the individual star islands—the galaxies—would move away from each other with speeds that increase with the distance from the center. Thus, for the first time, the theoretical possibility of an expanding Universe was introduced into the mainstream of cosmological thinking.

In 1922 Georges Lemaître was ordained a priest in Belgium, but he continued his studies in mathematics and physics. In 1927 he obtained his Ph.D. from the Massachusetts Institute of Technology, and in that year his important theoretical work, "Expansion of the Universe" was published. Lemaître posited a special model of the Universe that was closed and contained a repulsive force slightly greater than that chosen by Einstein. Thus, Lemaître mathematically introduced an expansion because his repulsive force was slightly larger than the gravitational force. This led him to introduce the concept of a beginning from which the expansion started. Lemaître named this beginning the "Primeval Atom" or the "Cosmic Egg." Initially, he reasoned, the Universe must have been compressed to a state of very high density. Lemaître is justly called the father of the concept of the Big Bang.

In the scant ten years following publication of Einstein's paper on the first mathematical model for the Universe, great strides had been made in the theoretical development of the model. But, in contrast to earlier creation myths, this evolving twentieth-century "creation myth," or model, was growing out of scientific physical principles and the strict application of mathematical tools. Those modern models rested on a foundation of rational scientific principles unparalleled in earlier thinking on creation myths.

By the end of the 1920s, the theoreticians' mathematical model of the Universe assumed a smooth distribution of matter, which led to the prediction of an expanding Universe that had a beginning in a highly dense state.

As the theoretical cosmologists played around with their model universes, the observational astronomers began to make sense of some of the patterns of motion among the stars and galaxies. The analytical tool at their disposal was the spectroscopic Doppler effect. This effect is much like the change of pitch in the tone of an ambulance siren passing you on the street. When the ambulance approaches you, the pitch of the siren sound is higher; when it recedes, the pitch is lower. The same phenomenon occurs with light, which in a sense vibrates just as sound vibrates. But when a light source is approaching you, the color shifts toward the blue end of the spectrum, whereas when the light source moves away from you, the color shifts toward the red end. We speak of this phenomenon as a *redshift*.

In the late 1920s, this Doppler tool was applied systematically to stellar spectra to study the motions of stars around us. (A spectrum of a star is its original white light spread out in a band of colors by, for example, a prism.) By analyzing thousands of stellar spectra—and working independently of each other—Dutch astronomer Jan-Henrik Oort and Swedish astronomer Bertil Lindblad discovered in 1927 that our Milky Way galaxy is rotating. The Milky Way is like a spinning wheel that takes 250 million years to make one complete turn—an incredibly slow rotation. That discovery opened up a dynamic picture of the "nearby" part of our Universe. The stars that appear to be moving in a stream relative to our Sun, observed earlier in 1922 by Dutch astronomer Jacobus Cornelius Kapteyn, were found to reflect a general and very slow rotation of our galaxy about a center in the direction of the constellations Centaurus and Sagittarius. This center of rotation is located some 30,000 light years away from us. For comparison, aside from the Sun, the nearest star to us, Alpha Centauri, is only 4.3 light years away.

So a picture emerges of the Milky Way as our own star island with something like 200 billion stars, spinning slowly like a flattened wheel. During the last rotation of this gigantic star wheel, the dinosaurs came and went and mammal life appeared—including human beings. And during humanity's written history (5,000 years or so), this star-studded wheel has turned but a minute fraction of a revolution.

What about the motions of other "Milky Ways"? Whereas the European astronomers, equipped with relatively small telescopes, studied stellar motions within our galaxy, American astronomers began to study the motions and structures of other galaxies. They used what were then the largest telescopes in the world: the 60- and 100-inch-diameter mirror telescopes of Mt. Wilson in southern California and the 24-inch lens telescope at Lowell Observatory in Flagstaff, Arizona. With these telescopes, American astronomers began to study the structure of the larger Universe outside our own galaxy. Bear in mind that no one knew for sure what the scientists were observing when the studies first began in 1912. The American astronomers focused their attention on some apparently unknown fuzzy-looking objects in the sky. They had culled about 100 of these objects from a catalog prepared in 1781 by the French comet hunter Charles Messier. He had prepared his catalog of these immovable objects in the sky so as not to mistake them for the fuzzy-looking heads of comets.

Under the scrutiny of the more powerful telescopes, some of these fuzzy Messier objects were found to be relatively nearby members of our own galaxy. Some of those nearby objects turned out to be spherical clusterings of millions of stars, which the astronomers called globular clusters. Others were found to be distant looser clusterings—the open clusters of a few hundred stars. Some fuzzy Messier objects turned out to be genuine patches of bright, luminous interstellar gas.

About one-third of the Messier objects, however, could not be identified at all until just seventy years ago—pointing to the recentness of our own creation myth and the explosive growth in our understanding of the larger aspects of the Universe around us. When the American astronomers first trained their new and powerful telescopes on those unidentified Messier objects, they intended to measure the motions of those objects relative to us based on the Doppler effect. They also hoped, if possible, to unravel the objects' real structures. Were the objects simply nearby gas clouds, or were they distant star islands—other Milky Ways?

The easiest task was to measure the relative motions of the fuzzy Messier objects by evaluating them in terms of the Doppler effect. Did the objects approach us or recede from us? Vesto Melvin Slipher first systematically undertook that task with the Lowell Observatory telescope. In 1924 Slipher published an analysis of the spectra of forty-three fuzzy objects, many of them Messier objects. The overwhelming majority of those objects were moving away from us, and the fainter they were, the faster away from us they appeared to move. Assuming the degree of faintness of these fuzzy objects were a measure of distance, Slipher had laid the observational foundation for the relationship between a fuzzy object's distance from us and its velocity. *He provided the first observational evidence of an expanding Universe.*

However, the issue was not yet resolved. What if the faint fuzzy objects were not stellar groupings far away from us but instead just a peculiar type of nearby gas cloud? The clincher would obviously be to find evidence of *individual stars* within those fuzzy objects. This more difficult task was undertaken simultaneously with the Slipher study by another American astronomer, Edwin Hubble. Using the larger telescopes at Mt. Wilson, Hubble worked with a special group of stars that vary rhythmically in their brightness. These stars are called Cepheid variables, after the prototype Delta Cephei. American astronomer Henrietta Leavitt had discovered in 1912 that these variable stars obeyed a very useful relationship called the period-luminosity relation. From a study of the time it takes for the brightness variation to repeat itself, one can deduce the true brightness of the star. Knowing this, one can deduce the distance to the star by observing how faint the star appears to be.

Hubble used this tool to search for Cepheid variables in some of the objects that Slipher observed in his study of the fuzzy Messier objects. By actually finding Cepheid variables in these objects, Hubble proved that the Messier objects were indeed groupings of stars and not nebulous gas clouds. And by measuring the periods of the Cepheid variables, Hubble was able to determine the distance of the fuzzy objects. He found that they were all located far outside our Milky Way, within their own separate galaxies several million light years away. Our nearest neighbor, the Andromeda galaxy— which, incidentally, is visible to the naked eye—is located about 2 million light years away. The light from that galaxy that now reaches our eyes when we look to the sky started its journey at the time one of our first distant human ancestors, "Lucy," walked across the African volcanic landscape.

By 1929 Hubble had determined the distances to enough of Slipher's fuzzy objects that he could firmly establish the observational relationship between velocity and distance for galaxies outside our own. He found that the farther away another galaxy is from ours, the faster it moves away from us.

Thus, the painstaking efforts of Slipher and Hubble over many years firmly established an observational basis for the notion of an expanding Universe. The expansion velocity of a particular galaxy outside our own is equal to its distance from us multiplied by a constant known as the Hubble constant. Current estimates put this constant at 15 kilometers per second per million light years. An object at a distance of 2 billion (or 2,000 million) light years would move away from us at a speed of 15 × 2,000 = 30,000 kilometers per second, or one-tenth the speed of light.

If we assume that the observed expansion velocity is constant, we can calculate the time elapsed since the expansion began from a point source. This time is called the Hubble period, and it is currently given the value of about 15 billion years. Though there are some reservations concerning the assumptions and observations involved, the Hubble period is considered a measure of the age of the Universe.

Such was the state our modern creation myth had reached with the findings of Hubble and Slipher. Some 15 billion years ago, a monstrously powerful explosion took place within a very small space. From this explosion millions of galaxies rushed out in all possible directions. And yet, with millions of stars in these millions of galaxies, space was basically empty of matter. Humanity was reduced to sitting on its tiny planet—literally a cosmic grain of dust—swirling around a common star and located in the outskirts of the Milky Way, far away from any center of action.

However, in contemplating this vast Universe with modern observational tools, humanity for the first time had constructed a creation myth out of rational mathematical and observational thinking. Theoreticians like Einstein, de Sitter, Lemaître, and others had provided theoretical frameworks for a Big Bang scenario concerning the creation of the Universe. And marvelously enough, this theoretical scenario, which predicted an ongoing expansion of the Universe, was apparently confirmed by the observational evidence provided by astronomers.

If the new myth were true, then astronomers should be able to uncover evidence of the Big Bang itself. Such astonishing evidence would be forthcoming in the decades after the end of World War II in 1945.

But before considering an account of those events, we must make a detour via nuclear physics and sunshine. Our century has witnessed many stupendous resolutions of Nature's riddles, which would have been undreamt of a mere 100 years ago. Who in the late 1800s would have believed that only a generation later we would be able to explain quantitatively why the Sun and the stars *shine*—that nuclear reactions make them pour forth prodigious amounts of energy in the form of light?

In 1939, working independently of one another, German physicist Carl Friedrich von Weiszäcker and American physicist Hans Bethe used laboratory data and theory to construct theoretically a chain of nuclear reactions that could take place in stellar interiors, thus explaining the generation of energy that produced sunlight and starlight. Nature, with the greatest of ease, achieves in stellar central cores (at temperatures of 10 to 20 million degrees centigrade) the immensely powerful process of *fusion* that we now spend millions of dollars trying to imitate.

In the Bethe-Weiszäcker stellar fusion theory, the elements *carbon, nitrogen,* and *oxygen* participate in an essential manner in the nuclear reactions. But how were these elements created in the first place? Although most of us live under the impression that the atomic constituents of matter have existed forever, this is not necessarily so. During the 1930s, nuclear physicists came to realize that all chemical elements heavier than hydrogen had atomic nuclei that were all built up of basically two elementary particles: the proton (a hydrogen atom's nucleus), with a *positive* charge; and the neutron, slightly heavier than the proton and with *no* electric charge. After World War II, some nuclear physicists began to speculate as to how elements heavier than hydrogen could have been created.

Seeking in part an answer to this problem, American-Russian physicist George Gamow, together with Americans Ralph Alpher and Robert Herman, published a pioneering paper in 1948 wherein they tried to explain the synthesis not only of helium but of all naturally occurring elements. Their efforts led them to consider the early stages of the Big Bang.

They thought the location of the Big Bang could be a possible site for such synthesis because, with such a huge concentration of matter, the central temperature would be not 10 million degrees centigrade, as in the stellar interiors, but 1 *billion* degrees. By their calculations, under those remarkable temperature conditions, about 50 percent of the initial matter would be converted into helium. This would account for the formation of helium out of hydrogen in the early stages of the Big Bang, but Gamow, Alpher, and Herman could not find a way to "cook" (that is, to synthesize) the elements heavier than helium. The Big Bang fireball would expand so rapidly that the temperature and density would drop too fast for a subsequent fusion of helium into carbon and further synthesis into the heavier elements.

The solution to this dilemma was demonstrated in important papers published in 1952 by British astrophysicists Fred Hoyle and Geoffrey and Margaret Burbidge and American nuclear physicist William Fowler. These scientists were able to show conclusively, on the basis of laboratory data and theory, that massive stars in their later stages of development would produce suitable temperature and density conditions to synthesize *all* of the heavier elements. Concurrent analysis of the light spectra from the stars and distant galaxies has shown incontrovertibly that the abundance of all elements heavier than helium amounts to only 1 percent of the original hydrogen abundance.

Let us reflect on this knowledge in light of its significance for life on Earth. According to the astronomers' observational evidence, 99 percent of matter in the Universe is made up of the two lightest elements: hydrogen and helium. Life on planet Earth is made up mainly of the heavier elements, all of which have been created in stellar interior furnaces. Life—including us—is literally made up of star dust. With the exception of hydrogen, every

Also in 1965, however, at nearby Princeton University, American physicist Robert Dicke had been building a radio antenna to search for the radiation afterglow from the Big Bang. When he heard of the remarkable observations of Penzias and Wilson, he immediately put two and two together. The observational results of Penzias and Wilson and Dicke's new theoretical predictions were published simultaneously. Penzias and Wilson received the Nobel Prize in physics for the year 1978.

So, by the mid-1960s, our modern creation myth of the Big Bang rested on two undisputed observations: the cosmic helium abundance and the cosmic radio wave background. Both were strong observational evidence for the existence of an exceedingly hot and dense early stage of the Universe, and the observations stimulated a host of theoretical workers in the fields of nuclear physics, relativity theory, quantum theory, and cosmology to try to describe in detail the first early stages of the Big Bang event. As a result of more than twenty years of such efforts, a marvelous scenario has been constructed—in particular for *The First Three Minutes,* as described in the superbly written book of that title by Nobel Prize winner Stephen Weinberg.

It is only fair to mention that a variety of other exotic cosmological scenarios have been proposed in the course of the twentieth century. However, none can explain as satisfactorily as the Big Bang theory the observations of the cosmic radio wave background and the cosmic helium abundance. Most cosmologists today embrace the Big Bang theory as the most likely one, representing our modern creation myth of the Universe. They do not all agree on the details during the first second of the Big Bang (see Stephen Hawking's *A Brief History of Time*), but they do agree pretty much on the details after that point.

One of the clearest, most evocative and illuminating accounts of this scenario from the very first beginning to the present stage of the Big Bang has been provided by British-American nuclear physicist (turned astrophysicist) E. R. Harrison in his book *Cosmology*. This book is a treasure trove of scientific and historical information and is written in an extraordinarily articulate manner. The following account owes much to Harrison's version of this scenario.

Harrison prefers to move back in time from the present moment and thus to run the Big Bang scenario backward. From our present vantage point in space and time, we see by and large all galaxies moving away from us and from each other. We observe that those galaxies farther away from us move away faster.

As we move back in time, the galaxies will obviously become more and more densely packed together (see Figure 1.1). If we assume that the presently observed velocities have been the same all along, there would have been a moment when the galaxies almost touched each other. Astronomers currently specify that time as some 10 to 20 billion years ago. This is the time it would take to compress the matter of the observable Universe to a density high enough for galaxies to be formed. This epoch represents what Harrison calls "the heroic age of galaxy formation." According to the theoreticians, this critical density would be reached after the Universe was some 100 million to 1 billion years old. (Our own solar system formed much later, when the Universe was 5 to 10 billion years old.)

Let us move further back in time to when the observable Universe was further compressed and only 10 million years old. This "prenatal" era of galaxies has, because of the compression, warmed up to room temperature, or about 20 degrees centigrade.

In 1992 exciting new observations over the entire sky of the intensity of the cosmic radio wave background were reported by a team of American scientists from the Cosmic Background Explorer (COBE) project. This team observed the cosmic radio wave background emission to a much higher precision than Penzias and Wilson had done. The COBE team was able to demonstrate unequivocally that there were variations in the intensity of the cosmic background radiation. This provided much-needed evidence for the *clumpiness* of the associated matter, indicating that the matter distribution in this very early stage of galaxies was *not* completely smooth. The clumps provided pockets of higher-density matter that could form the "seed-spots" for matter to accumulate and eventually form galaxies. Some scientists have exclaimed that the "Holy Grail" of cosmology has been found! Whether or not that proves to be the case, undoubtedly the COBE observations have provided a very important piece of observational evidence to substantiate the Big Bang theory.

At still an earlier epoch—when the Universe was 1 million years old—the primeval matter began to glow at the temperature of about 700 degrees centigrade. Then, as we go back to between 100,000 and 1 million years after the Big Bang, the Universe becomes flooded with light at a temperature of over 3,000 degrees centigrade. We have moved back in time to the light-dominated stage of the Big Bang, when the Universe appears as a cosmic fireball, only a few feet in diameter.[1]

As we journey back to when the Universe was only 100,000 years old, we enter the splendiferous phases of the actual Big Bang. The scenario takes on the qualities of *creation*. Thus, the entire time interval from 100,000 years to, say, 15 billion years is only the aftermath of this creation epoch. Our modern creation myth envisions a stage for the existence of this cosmic fireball from the end of the first second to the end of the first million years. This is the time span for the Big Bang to exist as a Big Bang. After that, the debris of the initial explosion—light and matter—separate. An independently expanding matter cloud exists out of which the galaxies form, as well as an independently expanding cloud of light that Penzias and Wilson will discover as an afterglow in 1965.

The main constituent of the cosmic fireball, especially in the very early stages, is electromagnetic radiation. The energy, and thus density, of this light is incredibly high: one-tenth of a ton per thimbleful. The temperature is a colossal 10 billion degrees centigrade, a thousand times hotter than in the central cores of stars. There is still matter in this fireball but at a much lower density than that of light. In fact, the cosmological theoreticians predicted that at the age of one second for the cosmic fireball, for every particle of matter (hydrogen atomic nuclei) there would exist 1 billion particles of light, or

---

[1]If matter has been propelled from the "hazelnut" for 1 million years, why, at this point, would the fireball-Universe still be so small? It has to do with the enormous gravity holding back the powerful outward thrust of the high-temperature energy field. The Big Bang was not a conventional explosion.

Figure 1.1

THE BIG BANG

The development of our Universe over time – our modern creation myth

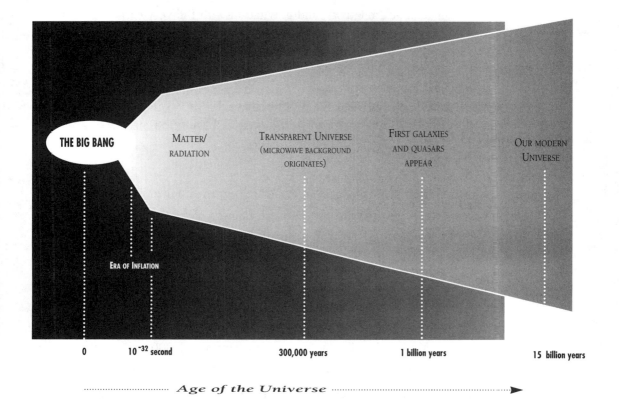

The Big Bang starts at time zero, its real size vastly smaller than a dot. Primeval helium is created by hydrogen fusion during the next 3 minutes. The cosmic microwave radiation separates from matter clouds 300,000 years later. The first galaxies appear 1 billion years later. Our modern-day Universe appears approximately 15 billion years after the initial explosion.

photons (see E. R. Harrison, *Cosmology,* p. 350). This prediction is verified today by observations of the intensity of the cosmic radio wave background.

What happened *before* the first second, before the commencement of the Radiation Era, as it is called? Those readers who wish to pursue this admittedly complex part of our modern creation myth can do so in Stephen Weinberg's *The First Three Minutes.* This part of our creation myth unifies, in a most beautiful and exciting manner, deep characteristics of nuclear physics, quantum mechanics, relativity, and gravitation with the very earliest stages of the Universe—all in incredibly small time scales during the very first second. This part of our scenario weds the ideas of the physics of the incredibly *small* to the physics of the colossally *large.* The scientific research in creating that unification must have been an extraordinary experience for the scientists involved, on par with first viewing Michelangelo's Sistine Chapel or hearing Bach's Toccata and Fugue in D Minor.

Where does this leave us in our search for the intellectual material to erect a new worldview? The astronomer's observational analysis of the abundances of the chemical elements in the Universe tells us that the particular constituents of matter that create the organic life forms—carbon, nitrogen, oxygen—are themselves cosmic debris, the ashes of burned-out stars. The stuff of which living species are created is itself only a minor constituent of matter. For every carbon, nitrogen, and oxygen atom in the Universe, there are roughly 10,000 hydrogen atoms and 1,000 helium atoms. For every magnesium, calcium, and iron atom, there exist 100,000 hydrogen atoms and 10,000 helium atoms. Indeed, the stuff of life itself is a second-order minor constituent of the Universe. These are hard, incontrovertible facts based on highly accurate present-day observations.

No matter what scenario we envision for the details of the creation of the Universe, we are left with the incontrovertible observations that a flood of light dominates our Universe today and that matter is numerically insignificant in the stuff of the Universe. For every particle of matter there are 1 *billion* particles of light. As Nobel Prize winner Ilya Prigogine puts it, matter is just a minor pollutant in a Universe made of light. Perhaps there is more to the phenomenon of light than meets the eye of the physicist. At the moment of almost ultimate compression, this light of the whole Universe was able to fit into a volume much smaller than the point of a needle.

The prophetic words of the great poet and mystic William Blake come to mind:

To see a world in a grain of sand
and heaven in a wild flower
Hold infinity in the palm of your hand
and eternity in an hour.

In this chapter we have followed the emergence of our modern scientific creation myth for the Universe. According to that myth, we human beings are inextricably connected to the very fabric of this vast Cosmos in at least two ways.

First, every single hydrogen atom in our bodies was once part of the Big Bang itself. Over 60 percent of the atoms in our material bodies were once inside the flaming inferno of the cosmic fireball. Furthermore, all the atoms heavier than hydrogen, such as iron, magnesium, oxygen, and carbon, were once forged out of hydrogen in the stellar furnaces of older stars. Forty percent of the atoms in our material bodies are recycled star dust.

This cosmic connectedness is shared by all life forms on this planet, be they horses, flowers, fish, or insects. It is also shared by rocks and stones and water and air—all the nonliving matter. The next step in our quest for understanding will thus concern not the creation of the Universe but the creation of our own planet Earth and the life it sustains.

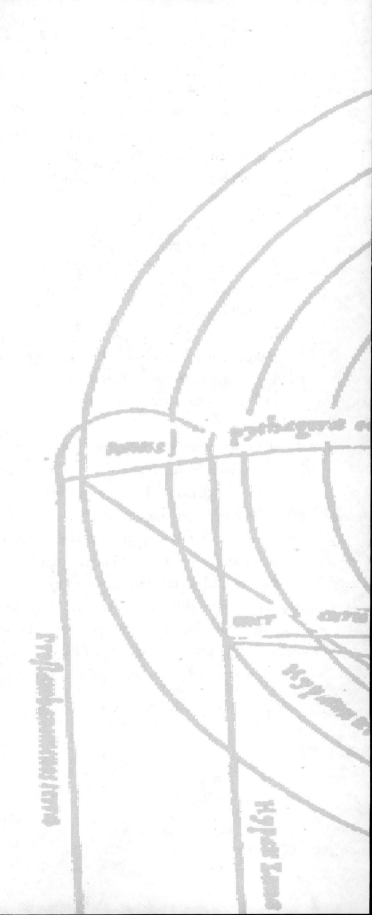

The FIRST STIRRINGS of LIFE

I n round numbers the history of our own solar system—the creation of
our sun and the birth of our planet—spans about 5 billion years. Let us
assume that the "Chronicles of Life on Earth" (or the chronicles of our
solar system) encompass a book of 1,000 pages. Each page of these chronicles, then, encompasses 5 million years of history.

It is probably quite significant that creation myths arise in the context of
every civilization that has tried to answer the eternal questions: Who am I?
From where do I come? To where do I go? Modern Western civilization is no
exception. In contrast to creation myths of earlier civilizations, however, the
modern version is based on numerous detailed observations of the material
constituents of our solar system coupled with theoretical calculations.

Amazingly enough, only in the last fifteen to twenty years has a coherent, self-consistent, and detailed scenario for the birth of our own solar system emerged. It is highly probable that this modern version is here to stay
and that further theories will only be elaborations upon it. This *Accretion
Theory of Planet Formation,* as it is called, is the composite work of many
modern scientists, among them American physicist Harold Urey and

Swedish physicist and astrophysicist Hannes Alfvén, both Nobel Prize winners in physics.

The modern scenario has as its starting point the Kant-Laplace theory of a contracting and rotating cloud, first described in the eighteenth century. Elaborations on the modern details of this theory have involved a multitude of painstaking observations and theoretical calculations, a considerably more laborious process than for the Big Bang theory. Nonetheless, recent observations are tentatively beginning to confirm the validity of this theory of how the planets formed.

In Chapter 1 we spoke of the "heroic age of galaxy formation," some 15 to 20 billion years ago, only a hundred or a thousand million years after the Big Bang event. That is the epoch in which our own Milky Way was formed. At present, galaxy formation has stopped because the expansion of our Universe has diluted matter in space far too much for galaxies to coalesce. Hence, our own Milky Way represents a very old stage of the Universe. However, much of its present material of gas and dust has gone through "cosmic recycling" in several successive generations of star formation. Our Sun belongs to a relatively late generation of stars. If our Sun were among the first generation of stars to be formed, we would not exist. The reason for this, we now know, is that all the atoms heavier than hydrogen in our bodies were formed inside a star.

If it were a first-generation star within our Milky Way, the Sun and its accompanying planets would not contain any heavier elements, such as carbon, nitrogen, and oxygen, because there would have been no earlier-generation stars to produce these heavy elements. Consequently, the chemical base for creating organic life like us would not exist. Thus, it is a very lucky "accident" that our Sun, with its age of about 5 billion years, is a relative newcomer.

The Sun was born out of an interstellar cloud that had been enriched with heavy elements produced by several earlier generations of massive stars. Those massive stars ran through their life cycles in a few hundred million years, then "went supernova" and hurled their contents, full of synthesized heavier elements, out into the surrounding gaseous environment.

Thus, our modern scenario for the birth of our Earth starts with an interstellar cloud floating around in the Milky Way some 5 billion years ago. The cloud probably would have continued as such were it not for some nearby star that went nova or supernova, exploded, and threw its heavier elements outward in a rapidly expanding shock wave, which hit and altered our protosolar gas cloud.

As a result of this triggered breakup, our protosolar cloud began to contract and rotate. The contraction, a result of gravitation, pulled the matter inward toward the center of the cloud, and as the cloud spun, it was elongated by the centrifugal forces that pulled outward in a plane. Those combined motions resulted in the formation of a central rotating core (the protosun) and a rotating flattened disk (the protoplanets).

What material was our "mother cloud" made of? From modern observations we know that such interstellar clouds are made of three components: gas, dust, and a weak magnetic field. The gas is mostly composed of the two lightest chemical elements, hydrogen and helium, whereas the dust particles contain a variety of heavier elements like iron, aluminum, and silicon, the basic ingredients in ordinary stones.

From this configuration of a spinning core and a rotating disk come two separate phenomena leading to the creation of our solar system: the birth of our Sun and the birth of the planets.

The birth of our Sun is believed to have proceeded as follows. When matter in our protosun was compacted, it began to heat up. Eventually the temperature in the center became so high that the dust grains melted and evaporated into gases. As matter continued to fall into the core, the temperature continued to rise until it reached a few million degrees centigrade. At that high temperature, nuclear reactions could begin to take place in the core interior, and the real "sunshine"-producing mechanism was turned on. The fledgling Sun was born and would continue to shine for thousands of millions of years.

As the young Sun struggled to develop, gas and dust *outside* the rotating disk fell freely into the core to build up and add heat to the protosun. However, those same materials *within* the disk were kept from falling into the central core by the centrifugal force of the disk's spinning action.

Here and there in this disk, accidental whirlpools formed as a result of the turbulent internal boiling motions. More gas and dust were flung into those eddies, and small local centers of gravitational attraction formed. Those centers pulled in more gas and dust, which began to form clumps of solid matter. They grew in size, much like rolling snowballs. The sizes of those "cosmic snowballs," though, ranged from that of pebbles to mountains. The cosmic snowballs were *dirty*: besides frozen ices of the lighter elements (hydrogen, carbon, nitrogen, and oxygen), they also contained dust grains made up of the heavier elements (such as iron, aluminum, manganese, and uranium). Such clumps composed of the original matter of the "mother cloud" still exist today in the form of comets. That is why scientists hope eventually to land an astronaut on the surface of a comet—to dig up samples of that original material.

At this stage began the birth of the planets. In the midst of all the cosmic snowballs orbiting around the newborn Sun, jostling motions periodically made the clumps collide with each other, the smaller clumps being shattered and compressed into larger chunks. As those chunks grew more massive, their gravitational pull drew the smaller bodies into their ken to be swallowed up. Thus, those larger bodies grew into the size of the protoplanets, which continued to sweep up the debris of the smaller clumps in their almost circular orbits around the new Sun.

Space probes like Voyager I and II and the Mars, Mercury, and Venus probes have all shown that everywhere in our solar system the planets and their moons are pockmarked with craters. These craters are now believed to be the result of this steller debris being swept up—the result of collisions between thousands of planetesimals a few hundred kilometers long and the planets and moons. These events took place 4.6 billion years ago during a very brief span of a few hundred million years, perhaps most of it during the first 100 million years of the formation of the solar system.

Meanwhile, the struggling young Sun put out not only powerful blasts of radiation but also solar winds, which are energetic streams of particles that blow away much of the inner particle debris and outer gaseous envelopes of the terrestrial planets. During the few hundred million years of "mopping up" the original interstellar cloud, the Sun settled down to a stable output of energy radiation, and the planetary system looked much like what it is today.

The final pages of our scenario for the birth of our Earth concern the early history of our solid planet after it condensed from the jostling planetesimals in the protoplanetary cloud.

The collisions of the planetesimals with each other to create a proto-Earth led to heat being generated, which in turn melted the accumulated material. In the course of this melting process, the heavier elements sank to the center of the proto-Earth and formed a central core, while the lighter elements floated on top. From studies of the propagation of earthquake waves, geologists know that the Earth now possesses a large iron-nickel core, on top of which lies a mantle made of molten rock. At its surface, this rock forms a crust of solidified minerals containing mainly oxygen, silicon, and magnesium.

The central core is still heated by the radioactive decay of heavy elements like radium and uranium. That heat then drives convection currents in the molten interior, which in turn create the Earth's magnetic field. Those convection currents also move about the crustal plates, which form the continents and the ocean floors.

In the earliest history of solid Earth, however, there were no oceans, only an atmosphere that would be highly poisonous to living things, much like that found on Jupiter or Saturn today. The early terrestrial atmosphere was made up of hydrogen and helium but also of ammonia, methane, carbon dioxide, and water vapor. That primeval atmosphere constituted only a small fraction of the Earth's mass and contained virtually no free oxygen or ozone. The water vapor and carbon dioxide in the primeval atmosphere may have come from outgassing during heavy volcanic activities. Those gases had been trapped in the minerals of the rocky crust. With no ozone in the atmosphere, the surface of the young Earth was fully exposed to the Sun's lethal ultraviolet radiation. Such was the primitive Earth in the beginning of its life.

If we could have visited this young Earth as observers from outer space, we would have found a planet totally alien to the life forms we know today. Given the choice, we probably would have rejected it as an abode fit for life. Primitive Earth was hot and stuffy, replete with thunderstorms and a deluge of rains and lightning. However, life did arise, as we shall explore in the next few chapters. To follow that story of the unfolding of life on this seemingly totally inhospitable Earth is a tale far more fascinating than any science fiction story ever dreamed of. Some will say that it was all the result of chance. The thesis of this book is that it was not.

Until now, we have been concerned only with such relatively "simple" questions as the origin of the Universe and the origins of our planet. These modern scenarios are grounded in certain basic natural laws that govern the behavior of inorganic matter. This matter could have remained in its state of "deadness," obeying these same natural laws forever without any appearance of life. Yet, life did start on Earth some 4,000 million years ago. What are some of the current hypotheses for the origin of this life on our planet?

To resolve the question of what prompted life to evolve on Earth and to understand the literally millions of shapes that present and past life forms display, we shall eventually have to approach the issue of fusing mind and matter.

According to the Bible, God created the objects of inanimate matter first—Heaven and Earth—and then the world of living forms, culminating with the first humans, Eve and Adam. That story is but one of the many myths of *special* creation, wherein a sharp distinction is made between living (or organic) and nonliving (or inorganic) matter. Such an illusory duality persisted for most of our history, until modern chemistry was born only a century and a half ago. In 1828 German chemist Friedrich Wohler succeeded in creating a relatively simple organic substance out of *inorganic* material. That feat came as a shock to the biologists and natural philosophers of the time and in one fell swoop eradicated the former distinction between organic and inorganic material. Today, it is obvious that living forms are created out of the same atoms and molecules that go into the making of non-living forms. A stone is composed of the same types of atoms that go into the making of a human body. Although the chemical composition may be different, the atomic structures are built on the same model. What matters is the *organization* of the atoms and molecules.

The Greeks, who were the originators of modern Western science, addressed the recurrent appearance of life forms—dogs and cats, flowers and insects, horses and humans—as acts of *spontaneous* creation resulting from special creative urges. These life forms were vestigial remnants in organic matter from the original act of creation.

With the rise of the Renaissance and the baroque period, medical scientists like British doctor William Harvey (1578–1657), who in 1628 discovered the blood circulation system in human bodies, speculated that life forms might arise from seeds and eggs too small to be visible to the naked eye. Indeed, through the design of the first microscope, Dutch scientist Anton van Leeuwenhoek (1632–1723) discovered the existence of myriad tiny micro-organisms in drops of water, which to some appeared as a possible bridge between living and nonliving matter. French doctor of medicine Louis Pasteur, however, showed in 1884 that those micro-organisms were not spontaneously created when they were deprived of air and nutrient broths.

Speculations about life on other worlds came to the fore, particularly with the development of the new astronomical worldview in the seventeenth and eighteenth centuries. In 1698 Dutch astronomer and physicist Christiaan Huygens published *The Celestial World Discovered, or Conjectures Concerning the Inhabitants, Plants, and Productions of the Worlds in the Planets.* That book led to an emerging (and unproven) belief that life might have evolved elsewhere and propagated through space to "seed" planets like our Earth. This panspermia theory, as it was called, was fully developed by Swedish chemist Svante Arrhenius, who in 1908 published *Worlds in the Making,* wherein he specifically developed the idea that spores or bacteria might be transported through space by the action of starlight.

It is highly symptomatic of the intellectual state of our times that a number of scientists and humanists other than those grounded in biology grapple with the question of the origin of life and show an increasing dissatisfaction with the Darwinian concepts of chance and natural selection. One of the most recent dissenters is a highly respected astrophysicist, Sir Fred Hoyle, who in 1983 published *The Intelligent Universe,* in which he champions Arrhenius's notion that life came to Earth from elsewhere. But Hoyle adds the "shocking" hypothesis that life was created by an intelligence elsewhere in the Universe. The present book, however, asserts that this intelligence resides right here on Earth.

The advances of science in the twentieth century convincingly make the case for a biochemical origin of life, for which the boundary between "dead" and "living" matter is completely blurred. Beyond any reasonable doubt, the atoms of living matter are in no way different from those in dead matter—just consider dead trees versus wood or fossils versus sedimentary rocks. The great advocates of the biochemical origin of life were British biologist John Haldane and Russian biochemist Alexander Oparin, who wrote several books on the subject in the 1930s. The crucial issue is the difference in the organizational structures of atoms in living matter versus dead matter.

Accordingly, we must follow the misty scenario of what may have happened on the surface of Earth as the rudiments of life's material were created, up until the appearance of the first living cell. Fossilized cells, formed about 3.8 billion years ago, appear in sedimentary rocks in Greenland. For want of older fossils, this age is the fragmentary benchmark for dating the first appearance of cells. Hence, the protolife material evidently was formed sometime in the period between the crustal solidification of young Earth 4.6 billion years ago (as dated by the oldest rocks on Earth and the Moon) and the formation of those fossilized cells 3.8 billion years ago.

To guide us in our fumbling thinking on these matters, two groupings of modern evidence come from two totally disparate sources: a terrestrial laboratory and interstellar gas clouds.

In 1953 an American graduate student, Stanley Miller, at the suggestion of his professor, Harold Urey, at the University of Chicago performed an experiment that caused a major sensation in scientific circles. In a large glass vessel, they introduced gases that matched the assumed composition of the first early atmosphere of young Earth, such as methane, ammonia, carbon dioxide, and water vapor. They then exposed that mixture to electrical sparkings of high energy (60,000 volts) for days on end through electrodes placed on the surface of the glass vessel.

Even though they regarded some of the results as theoretically possible, it was still much to everyone's surprise to find that a series of *organic* substances were formed and deposited in the water at the bottom of the jar—organic material formed out of the *inorganic* initial constituents. Sugars, fats, alcohols, and, most important, amino acids were formed quite spontaneously. Such amino acids are the fundamental entities that act as building blocks of proteins and enzymes, which in turn are vital to organic cell functioning.

It is all very well for such events to happen in a terrestrial laboratory under controlled experimental conditions. But are they really likely to take place in Nature? Favorable evidence has increasingly come from radio-astronomers' search for exotic molecules in interstellar gas clouds. Radio-astronomy is a novel branch of astronomy, developed by Americans Grote Reber and Karl Jansky in the late 1930s. Radio wave signals from the first interstellar molecule, hydroxide (OH), were detected in 1963. Since then the discoveries of increasingly complex interstellar molecules have grown with the speed of an avalanche. Today, we count more than 100 interstellar molecules, many of them organic in nature. But no amino acids have yet been detected in the interstellar gas clouds.

Nonetheless, on the basis of those two lines of evidence for the "natural" formation of organic compounds out of inorganic material, scientists now visualize (tentatively, but with increasing reassurance) the following scenario for the origin of the first life material before the cell was created.

The period in Earth's history immediately following the planet's coalescence into one solid body must have been one of intense volcanic activity over millions of years. The primeval atmosphere became increasingly dense by the continual outgassing of the frozen gases in the interior, which had been liberated from the icy, chunky planetesimals and then heated to melting and evaporation inside the hot Earth. From the outside, powerful ultraviolet radiation from the young Sun bombarded the primeval atmosphere. No ozone existed to shield the atmosphere. Incessant lightning storms flashed illuminations on the bleak, crusty surface where the oceans had not yet formed. Above all, Earth was searingly hot and continually racked by quakes.

As water vapor outgassed from the interior, however, the atmosphere eventually became supersaturated with water molecules. It began to rain, rain, for millions of years. In the beginning there was a continual recycling of the water, because the shallow lakes, ponds, and seas were very near the boiling point and rapidly gave the water back in vapor form to the atmosphere. During this period copious amounts of very primitive organic material were first forming in the atmosphere. They followed the rain down, partly to be destroyed again, partly to be reformed. Occasionally, a large planetesimal would collide with Earth and disrupt the whole process.

Finally, about 4.2 billion years ago, the turbulent birth process of Earth's surface terminated. The Earth had survived the epoch of planetesimal bombardment and had cooled enough to allow a more stable environment for protolife on the surface. The oceans were formed in final volume as the last outgassing took place. Today, the oceans cover about 70 percent of Earth's surface and represent about 0.03 percent of Earth's mass. The geometrical *shape* of the ocean coastal contours has varied incessantly as the continental landmasses have come and gone, swirled together and broken up by the motion of the crustal plates known as the tectonics. However, the ocean's *volume* appears to have stayed remarkably constant.

The large, dominant surface body of water obviously played an enormous role for the development of primitive life. The ocean effectively shielded the protolife components—the amino acids, sugars, fats, and nucleotide bases—from the destructive effects of the solar ultraviolet radiation. The ocean also acted as a temperature stabilizer, because water has the capacity to retain and distribute heat. Furthermore, the water vapor trapped the infrared radiation released from the ground by the solar radiation's direct heating.

Thus, a stable environment ensued that vitally furthered the development of protolife. Still, at this epoch of Earth's history, there were absolutely no surviving protolife elements on the barren land, and the atmosphere was highly poisonous, containing hardly any oxygen.

It was in the oceans that the "mystical" life processes took place during the next 3.5 billion years. During 85 percent of Earth's history, all landmasses have been totally barren. There were absolutely no flowers, trees, birds, insects, or other animals. During almost 50 percent of Earth's history, the atmosphere was highly poisonous to the oxygen-breathing type of life. Life on land thus represents a late stage in evolution. It was in the ocean that patient preparations were made to ready life for an advance onto land. Perhaps whatever guided the development of life in the oceans began to consider moving life onto dry land comparatively late. Numerous biological developments had to be readied to equip the life forms that would embark on a voyage to the barren landmasses.

The spectacular developments in microbiology and molecular biology during the last twenty years have enabled scientists to scrutinize the details of the organization of atomic structures in life forms. We are still far from seeing the end of the impact of molecular biology upon both science and humanism. The birth of modern molecular biology took place in 1953, when British biochemists Francis Crick and James Watson discovered the double-helix structure of the DNA molecules that constitute the chromosomes of cells.[1] Since that time, a thrilling scenario has evolved for the creation of life forms on Earth, a scenario with many chapters still remaining to be written.

We may think of the double-helix structure of DNA as the alphabet with which to begin reading the instructions for making life forms. In order for us truly to appreciate the elegance and ingenuity of the construction of this genetic code, we need to go into some detail of DNA's fine structure.

---

[1] It is worth noting that 1953 is the same year Miller and Urey announced their creation of organic substances out of inorganic matter.

Figure 2.1 (*left*) shows the double-helix coil, where one turn measures the incredibly small distance of thirty-four times the diameter of the hydrogen atom. The backbone of each coil is a long string of molecules called sugar-phosphates. Molecular biologists have adapted a simple diagrammatic scheme from chemistry that shows how the different atoms link up to form one of these molecules. Figure 2.1 (*right*) shows such a diagram for the string of the sugar-phosphate molecules forming the backbone of the DNA coil. The phosphorus atom (P) hooks up with four oxygen atoms (O). One of the oxygen atoms surrounding P also hooks up with CH_2, a molecule composed of carbon (C) and hydrogen (H), which in turn connects to a five-sided complex of atoms—a sugar molecule. The technical name for this sugar molecule is deoxyribose, which is made up of five carbon atoms, ten hydrogen atoms, and four oxygen atoms.

A DNA backbone element is thus made up of twenty-seven different atoms of four different chemical elements: H, C, O, and P. This backbone element repeats itself again and again along the DNA coil, much as the individual disks in our spines.

From Figure 2.1 we can see that the two DNA coils are hooked up to each other by ladder-type "rungs," labeled A–T, G–C, and so on. The letters A, T, G, and C stand for the organic molecules adenine, thymine, guanine, and cytosine, which are the only four letters of the DNA genetic code alphabet. These organic molecules are called nucleotide bases. Some hydrogen atoms in one molecule are lined up with other atoms in another molecule (for example, thymine and adenine) so that A and T always lock together, as do G and C. Thus, two coils of DNA are braided together to form what is called a chromosome.

Such details of the fine structure of the DNA genetic code are important to imprint upon our minds the ingenious and elegant chemical complexity of this life-giving structure. However, this is only the beginning of the fascinating tale of DNA. The next step is to understand that the manner in which A, T, G, and C are *sequenced* along one coil of the DNA helix gives instructions for the creation of organic molecules to be used in the assembly of life forms.

It turns out that one group of organic compounds that Miller and Urey created in the laboratory, the amino acids, are intimately connected to Crick and Watson's genetic alphabet. Here we have a most exciting coupling of two totally independent discoveries, something of vital significance for our emerging understanding of the creation of primitive life.

An example of an amino acid is alanine, $C_3H_7NO_2$. The genetic code for alanine turns out to be a three-letter "word" (or codon), which can be either GCA, GCC, GCG, or GCT. Whenever one of these sequences is found along one coil (arm) of the DNA helix, the amino acid alanine shall enter into the structure at that point of the larger life molecule (that is, the protein) being created.

Thus, one function of the DNA genetic messages written into the chromosomes is to spell out the creation of long linkages of amino acids, usually several hundred of them. These molecular chains, called proteins, are the vital building blocks for creating cells and their functions. Basically, only twenty amino acids are used by nature for this operation. Specific three-letter codons in the DNA structure are associated with each of these amino acids (see Table 2.1). In the next chapter, we will take up this protein-building process in more detail.

It is quite significant that only four chemical elements are involved in the formation of the genetic letters and their associated amino acids and that these four elements are hydrogen, carbon, nitrogen, and oxygen. Besides helium, which does not enter into the building of life forms, these are the most abundant elements in the cosmos. Furthermore, carbon, with its particular structure of six electrons around its nucleus, is chemically one of the most active elements. An average-sized protein is typically made up of 100 or more amino acids strung together in a variegated manner out of the 20 amino acids listed in the genetic code. About 200,000 different proteins make up existing life forms. In our bodies there may be more than 100,000 different proteins performing a variety of tasks, from making up bones, skin, and hair to carrying substances from one part of the body to another.

To correctly assemble amino acids into the various kinds of proteins, certain mechanisms must be able to read the genetic code instructions. It is

The
D N A
H E L I X

Figure 2.1

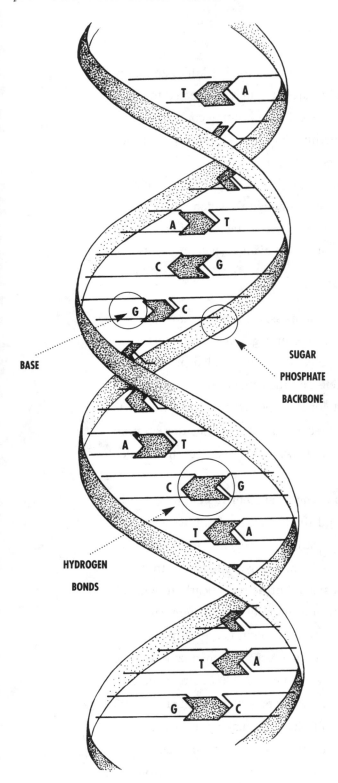

BASE

SUGAR
PHOSPHATE
BACKBONE

HYDROGEN
BONDS

In a human cell, the total length of the DNA double helix (that is, the aggregate length of 46 chromosomes) is 6 feet; it contains some 50,000 genes. Yet it is coiled together into a marvelously compact form—a small lump the size of a pinpoint. Microbiologists estimate that the total information stored within the human DNA is the equivalent of over 1,000 books. How is this information stored? In order better to understand the storage mechanism, we have to examine the molecular fine structure of the DNA helix depicted in

A schematic structural representation of the DNA double helix, which is the genetic constituent in the chromosomes of all cells. The length of one turn of the spiral is about 34 diameters of a hydrogen atom—a very, very small distance.

The SUGAR PHOSPHATE BACKBONE

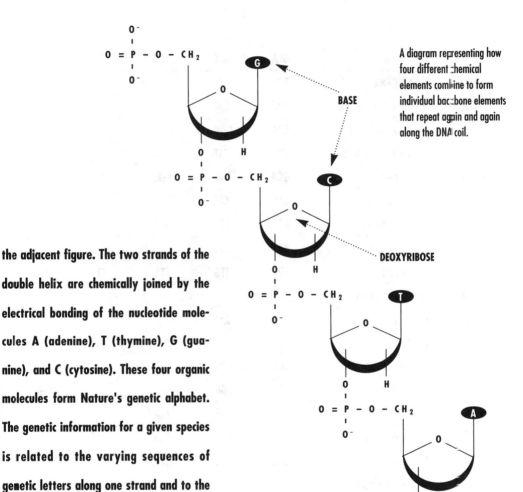

A diagram representing how four different chemical elements combine to form individual backbone elements that repeat again and again along the DNA coil.

the adjacent figure. The two strands of the double helix are chemically joined by the electrical bonding of the nucleotide molecules A (adenine), T (thymine), G (guanine), and C (cytosine). These four organic molecules form Nature's genetic alphabet. The genetic information for a given species is related to the varying sequences of genetic letters along one strand and to the interplay among different gene complexes.

THE GENETIC CODE

Amino acids and their codons

Table 2.1

Ala	ALANINE	GCA	GCC	GCG	GCT		
Cys	CYSTEINE	TGC	TGT				
Asp	ASPARTIC ACID	GAC	GAT				
Glu	GLUTAMIC ACID	GAA	GAG				
Phe	PHENYLALANINE	TTC	TTT				
Gly	GLYCINE	GGA	GGC	GGG	GGT		
His	HISTIDINE	CAC	CAT				
Ile	ISOLEUCINE	ATA	ATC	ATT			
Lys	LYSINE	AAA	AAG				
Leu	LEUCINE	TTA	TTG	CTA	CTC	CTG	CTT
Met	METHIONINE	ATG					
Asn	ASPARAGINE	AAC	AAT				
Pro	PROLINE	CCA	CCC	CCG	CCT		
Gin	GLUTAMINE	CAA	CAG				
Arg	ARGININE	AGA	AGG	CGA	CGC	CGG	CGT
Ser	SERINE	AGC	AGT	TCA	TCC	TCG	TCT
Thr	THREONINE	ACA	ACC	ACG	ACT		
Val	VALINE	GTA	GTC	GTG	GTT		
Trp	TRYPTOPHAN	TGG					
Tyr	TYROSINE	TAC	TAT				

This table represents the genetic code names (codons) for each of the twenty amino acids that enter into the manufacturing process of proteins in the cells.

During protein synthesis, whenever the synthesizing agent encounters a given codon (for example, GCA) along the DNA helix, the corresponding amino acid (alanine for this example) will be added to the protein segment being manufactured. The letter sequence TAA acts as a signal to end the synthesis of the particular protein.

exceedingly important that the linking of amino acids be done 100 percent correctly. One single mistaken placement of one amino acid can have disastrous consequences for the life form that uses the protein. A classic example is the case of sickle cell disease found frequently in Africa. It results from the faulty placement of just *one* amino acid of the more than 600 that go into making the hemoglobin molecule. This single misplaced amino acid leads to chronic anemia, jaundice, severe pains, and poor resistance to disease.

Although amino acids are comparatively simple combinations of anywhere from nine to twenty-four atoms, it is obvious that proteins are far more complex structures that must somehow have formed, albeit rudimentarily, in the "primordial soup." Likewise, proto-enzymes, special types of proteins that initiate or speed up vital chemical reactions in living cells, must have formed. In fact, fully developed enzymes such as we see in bacteria, the oldest life forms, must have been completely created at the time the first single-celled life forms were created some 3.6 billion years ago. By that time the "invention" of the cell membrane must have been complete—along with the "invention" of DNA having instructions for creating new bacteria. Thus, the phenomenon of cell replication was invented. And the time interval during which those inventions must have been accomplished—the time between when the Earth's oceans formed and the first cells appeared—is only 400 million years. That is only one-tenth the total age of Earth. Over the span of a few hundred million years, an astonishing amount of complex organizational structure was imposed on the "dead" matter of the atoms.

Many molecular biologists (Alberts et al. 1989, p. 10) believe that a simpler precursor of DNA was first "accidentally" formed in the primitive chemical sea. It is supposed to have been a single-stranded helix (not double as in DNA) with a simpler form of the sugar-phosphate backbone (ribose rather than deoxyribose). Finally, instead of the genetic code letter T (for the thymine molecule), it contained the code letter U (for the uracil molecule). A strand having this structure is referred to as RNA. Somehow DNA is supposed to have been formed as a "variation" on RNA. In the following chapter we shall consider the vital role RNA plays in protein synthesis.

So how did these highly ordered RNA structures evolve out of the primordial soup? How could the DNA coil strands evolve with their ladderlike rungs of the genetic code letters A, T, C, and G? And last but not least, *how could meaningful instruction sequences of these genetic code letters be put together?*

From this point on, accounting for the whys and hows of the increasingly complex molecular organization in the evolving life forms poses a major problem. It becomes increasingly difficult to describe the history of life on Earth in terms of the natural physical laws that apply to "dead" matter.

Conventional wisdom has it that there is no need to add any new laws or principles to describe the appearance of complex organic molecules such as proteins or enzymes. Conventional wisdom has it that these life molecules were produced by *chance* and *natural selection* (the Darwinian paradigm) at the beginning of the existence of oceans on our planet. But this Darwinian thought model has not been *mathematically* applied to the physical situation we have described, and for good reason. The standard scientific criticism made by nonbiological scientists can be illustrated by the following digression into the laws of probabilities:

Suppose we wish to create an early primitive enzyme, made up of only thirty-four amino acids. Remember that the DNA furnishes instructions on how to make life molecules, such as this specific enzyme, using only the four genetic letters A, T, C, and G. The enzyme can be made only by a very specific sequence of these letters; suppose this sequence starts with the letters AAGCCT . . . and so on. We shall need $3 \times 34 = 102$ correctly placed molecular letters in the DNA, three for each of the thirty-four amino acids.

The odds that we will pick the first letter correctly are 1 in 4 (or $1/4$) because there are four letters to choose from. The chance of picking two letters correctly in the sequence is not the sum of the two ($1/4 + 1/4$) but (according to the laws of probabilities) the *product* of the two, that is, $1/4 \times 1/4 = 1/16$.

We can now see that the odds of picking 102 letters in a correct, specified sequence are $1/4$ multiplied by itself 102 times, or as the mathematician would write it, $(1/4)^{102}$. So, according to the laws of chance, if we make 4^{102}

trials, one of them ought to be successful. It may in fact take more or fewer trials, but on average, 4^{102} trials should be enough to yield success.

In order to get a feeling for the time span involved in these consecutive trials, let us assume that each trial takes one second. That gives us 10^{61} seconds for the time it will take us to go through all the trials to get the correct instruction for the DNA for making this enzyme.

One year is made up of roughly 30 million, or 3×10^7, seconds. This means that our 10^{61} tries—at the rate of one try per second—will be accomplished in about 3×10^{53} years, or a time span vastly larger than the presently estimated age of the Universe, which is somewhere between 1.5×10^{10} and 2×10^{10} years. Even if we allow for a large number of places on Earth for these trials to be taking place simultaneously and shorter times for the trial, the chance argument of the Darwinists does not hold up mathematically. (Chapter 7 gives additional details on the mathematical inadequacies of Darwin's concepts.)

Of course, our example mathematically considers only the problem of linking by chance the equivalent of thirty-four bits of information into the DNA. The difficulties for the proponents of chance theory are vastly exacerbated by the fact that in the first single cells, which appeared about 3.8 billion years ago, the genetic information packed into the one double helix of DNA was already of the order of *1 million* bits of information—the equivalent of 125,000 words, or a 500-page book with 250 words of instructions each page.

Today we are all more or less familiar with the wonders of miniaturization in the field of modern computers. In particular, the smallness of computer memory chips is touted as an example of our superb high-technology achievements. The best effort nowadays is to pack 1 million bits of information within an area half the size of a postage stamp. However impressive this may appear, it is nothing compared with the achievement of Mother Nature. She packs that information on an area smaller than the tip of a pin—not the head of a pin, but the *tip*. This is the information stored in DNA to make a simple single-celled bacterium. In modern cells of humans, Nature packs an additional *1,000* times more information into that same area.

I n this chapter we have thus learned that within the cosmically relatively short time span of 400 million years, crucially significant events took place in the ocean or in coastal tide pools—the formation of RNA and the even more important subsequent creation of DNA.

In particular, the DNA structure was to become *the* repository for the information accumulated during future eons of biological design and evolution. Unvaryingly it has been the informational databank and program memory used in creating all life forms on this planet, past and present. The creation of DNA and its incorporation into a cell structure was the first and most important evolutionary watershed in the history of life on Earth.

The FIRST PRIMITIVE CELLS

For two-thirds of Earth's history, only single-celled life existed on the planet. There were no fishes, no marine life as we know it, no birds, insects, plant life, or mammals. Land was totally barren. Even the ocean harbored only single-celled life—seemingly a monotonous uniformity.

But as with the extensive preparatory stages before the first cell appeared, the long gestation period during which life dwelt in single-celled form involved a number of preparatory steps for the birth of multicellular life. Our next window of knowledge covers a time span of about 3.1 billion years—from the appearance of the first sedimentary fossil cells 3.8 billion years ago until the appearance of the first multicellular life forms about 700 million years ago.

The use of the word *cell* to represent the smallest form of life is credited to British natural philosopher Robert Hooke (1635–1702). He was an ardent, combative rival of Sir Isaac Newton in the late seventeenth and early eighteenth centuries. In 1665, looking at a sliver of cork substance through the microscope invented by Dutch clothmaker Anton van Leeuwenhoek,

Hooke discovered a number of "pores" or "cells," which he believed served as containers for the "noble juices" of plant life.

About 150 years later, German botanist Mathias Schleiden and zoologist Theodor Schwann were the first to develop a unified cell theory, which stated that all animal and plant life sprung from the same cell architecture; the building blocks were only put together in different ways. That remarkable conceptual breakthrough was further enhanced by a discovery twenty years later by German scientist Rudolf Virchow that most diseases arise in the cell.

Another 100 years passed before two research tools of immense analytic value to the microbiologist were perfected: (1) the electron microscope, invented in the 1930s, which permitted magnifications up to 1 million times for peering into cell structures, and (2) the ultracentrifuge, developed by Swedish scientist The Svedberg, which permitted cell constituents to be separated and subsequently analyzed.

Those instruments ushered in a revolution in microbiology and in our understanding of the marvelous intricacies displayed in cell structures. Twenty-nine scientists have won Nobel Prizes in cellular and molecular biology since 1953, the year in which Crick and Watson discovered the genetic code. Thus, it is only during the last forty-five years, since World War II, that a coherent and structurally meaningful conception of cell structure has been developed.

Armed with this modern knowledge of the cell, we now can scrutinize the fossil records of past cell life and consider its probable evolution. The fossil records point to a most significant and fundamental division of cell structures into two distinctly different types: those that do not show a central nucleus, the prokaryotic cells (*pro* or *proto* means "before"; *karyon* means "nut kernel"), and those that have a distinct nucleus, the eukaryotic cells (*eu* means "well" or "true").

Only two groups of living organisms today belong to the prokaryotes: bacteria and blue-green algae. The blue-green algae are now classified as cyanobacteria. However, they are remarkably similar in their physiology to algae and plants (see L. Margulis and K. V. Schwartz, *Five Kingdoms,* p. 48).

All the rest of terrestrial life today as we know it—the green plants, the animals, the fungi, and the protistas such as amoebas and protozoans—belongs to the class of eukaryotes. The structure of the eukaryotic cells is in some way beneficial to and responsible for the remarkable diversity of life as we know it today on Earth. The simpler prokaryotic cell model appears to be an evolutionary "dead end" today. Even so, the fossil record shows that for a very long period in the history of Earth, the prokaryotic cells reigned supreme. In fact, the oldest fossil records point to the appearance of the first prokaryotic cells some 3.8 billion years ago. The eukaryotic cells appeared much later, when the terrestrial environment was proper for them to exist. The oldest fossil records of eukaryotic cells indicate that this much more complex cell structure cannot have appeared until about 1.4 billion years ago. So, for 2.4 billion years, almost half of Earth's life—the simple, unicellular, nonnucleated bacteria and blue-green algae—totally monopolized life on Earth.

Why?

The simple reason scientists have found lies in the fact that Earth's atmosphere did not contain enough oxygen for the eukaryotic cells to flourish. In contrast to prokaryotic cells, eukaryotic cells need oxygen to function; they are *aerobic* (oxygen-using) forms of life, whereas all the early bacteria were *anaerobic* (non-oxygen using). Today, visitors to the volcanic pools of Yellowstone National Park in Wyoming are able to witness life as it may have been in the beginning, some 3.8 billion years ago. The mineral-filled pools, particularly rich in sulphur, are a benevolent environment for the bacteria and blue-green algae that grow there into mats and leathery sheets. There, life flourishes without oxygen.

The primeval atmosphere did not contain oxygen. That chemical element was locked up in the stable molecules of water and carbon dioxide. Hence, the first life forms evolving out of the primordial soup had to flourish in an atmosphere that would be highly poisonous to almost all forms of life on Earth today.

Conversely, the ancient prokaryotic cells we observe today have sought out environments (such as volcanic pools) that protect them from what to them is poisonous: an oxygen atmosphere. Because Earth's atmosphere has been filled irrevocably with oxygen, further evolution of the prokaryotic life forms was stunted some 1.5 billion years ago when the present oxygen content (21 percent of the atmosphere) appeared to have settled.[1] Life's opting for an oxygen-rich atmosphere represents one of the watersheds in evolutionary history.

How did the oxygen enrichment come about? Two extremely important chemical developments took place in the early evolution of prokaryotic cells some 3.5 billion years ago, namely anaerobic and aerobic metabolism. The end result of metabolism (food processing) is to release energy within the cells to (1) drive their synthesis of new proteins and the construction of other cell materials and (2) maintain the individual cells' life processes.

This period also saw the development of the process of photosynthesis, which uses light to drive a considerably more efficient process of metabolism. First, anaerobic photosynthesis evolved in the green sulfur bacteria, to be followed a few hundred million years later by aerobic photosynthesis in purple bacteria (Alberts et al. 1989, p. 384). The end result of this last complex chain of chemical reactions, besides the production of energy molecules, is to release oxygen into the atmosphere. Thus began the oxygenation of our atmosphere.

The details of the chemical reactions involved in these processes are beyond the scope of our discussion. A fuller understanding of the complex details of photosynthesis was achieved only ten years ago. For their work on the three-dimensional structure of a photosynthetic reaction center in the bacterium *Rhodopseudomonos viridis,* J. Deisenhofer, R. Huber, and H. Michel received the 1988 Nobel Prize in chemistry.

We can legitimately question how the complex enzymes, chlorophyll, and other molecules could have been created and then combined to yield

---

[1]The prokaryotic cells continued to develop oxygen-resistant strains. However, these could not compete with the far more complex and versatile eukaryotic cells. Even though the oxygen-resistant prokaryotes—such as aerobic bacteria—abound today, they never advanced to form multicellular structures.

the complex of reactions that take place in metabolic cycles and photosynthesis. No detailed mathematical answers have been given by the Darwinists—or, for that matter, by anyone else—as to how this feat could be accomplished over a mere 800 million years.

The first primitive cells, the prokaryotes (0.001–0.01 millimeters in size), are 10 to 100 times smaller than the later, more advanced eukaryotic cells (0.01–0.1 millimeters in size). This difference implies that significantly fewer atoms are placed into prokaryotes and that these cells have less organizational complexity. Yet, it is important to note that these prokaryotic cells, taken by themselves, represent extraordinarily complex, organized structures of some 100 million atoms. When we begin to examine in detail the organizational and functional structures of these cells, we should keep in mind that already in these first primitive cells (bacteria and blue-green algae) there appears evidence for what could be interpreted as an intelligent, rational, logical blueprint for life.

Life as we know it today implies food intake for energy and growth, the digestion of this food, and the subsequent disposal of waste products, as well as the process of reproduction and the manufacture of proteins. Precisely the same guidelines are found to exist for the survival of the first primitive prokaryotic cells. Although the organizational structure of the first cells was fairly simple compared with that of the eukaryotes, it was of a sufficient complexity to enable them to survive in vast masses for several billion years.

The prokaryotic cell structure consists of a membrane that encloses strands of DNA, the genetic blueprint, loosely coiled in the cell fluid. In this fluid, protein factories called ribosomes manufacture building materials to repair or maintain certain structures in the cell membrane and to produce molecules for new cells. On the average a given cell contains about 70 percent water (by weight) and 26 percent large molecules such as proteins.

Just as the ideas of "eating," "digesting," and "waste disposal" are apparent in these cell structures, so is the idea of locomotion embodied in some rudimentary limbs—called flagella or pseudopods, depending on their structure—which protrude outward from the cell membranes of certain prokaryotes.

The three basic constituents of a cell, then, are the DNA, ribosomes, and the cell membrane, with flagella or pseudopods added for some prokaryotic species. When we analyze the raw materials that go into building these structures—and into building eukaryotic or multicellular organisms as well—the remarkable fact emerges that only four major families of small organic molecules are used: sugars, fatty acids, amino acids, and nucleotides. The sugars are the food molecules for the cell, and the fatty acids are components of cell membranes. The amino acids are the building blocks of proteins, and the nucleotides perform this same role for the DNA. It is truly astounding that from the lowliest protozoan to the human genius of Bach, the same four families of organic molecules nurture and maintain these unbelievably diverse life structures.

Let us take a look at the fine structures of these ancient life forms, the prokaryotic cells, keeping in mind the inordinate complexities of their designs. As we shall see, serious intellectual difficulties exist in trying to explain in detail how such designs could come into being.

The formation of a cell membrane derives from a remarkable chemical quality inherent in such fatty liquids as the organic oil molecule. The cell membrane consists primarily of two layers of fatty liquid molecules called phospholipids. In the small space between these two layers, the phospholipid molecules attract water molecules. However, the outer portion of the outer layer and the inner portion of the inner layer *repel* water molecules. This arrangement lets the cell keep its own fluid in and external fluids out.

The cell membrane is incredibly thin, only 0.00001 millimeter. The membrane is a flexible, pliable liquid in itself, and its plasticity is vital for the further sculpturing of life forms. Nevertheless, it is resilient and tough and serves admirably as a substance into which are embedded proteins, the aggregates of hundreds of amino acids.

The function of the cell membrane is not only to hold in the interior cell fluid, the DNA, and the ribosomes. It also serves the vital function of allowing nutrients, "food," into the cell interior to be used as an energy source and as raw material for protein manufacture by the ribosomes. In addition, waste material is expelled through the cell membrane.

In recent years molecular biologists have discovered two geometrical configurations of proteins that appear vital to the interaction between the cell interior and exterior. These proteins are either globular or helical in shape, and they penetrate the membrane to varying degrees.

The exact functioning of the globular and helix proteins is still under investigation by microbiologists. A variety of functions have been discovered for specific proteins. For example, the globular protein acts as a receptor for substances to be transported into or out of the cell, whereas the helix protein appears to serve as a transportation channel, that is, providing openings in the membrane through which individual atoms or molecules may pass.

Further transport inside the cell involves specialized protein "baskets," which begin to form in the cell membrane near where a specific globular protein has captured a nutrient molecule ready for transfer. This molecule gets stored in a basket, which then closes and detaches itself from the cell membrane. The basket then moves to a specific place in the interior cell fluid where the nutrients are stored in a larger basket enclosure, the endosome. (Each basket, incidentally, is made up of special protein molecules called clathrin. One clathrin molecule is composed of over 1,000 amino acids.)

By a process that is not yet well understood, the nutrients are then released into a digestive "stomach," the lysosome, where special enzymes dissolve the nutrients for use either as an energy source or as building materials.

Finally, the waste material from the endosome and the lysosome enters into similar baskets to be transported back to the cell membrane, where it is excreted in a manner similar to that by which the nutrient entered.

Here in the detailed atomic-molecular structure of the food and waste transportation mechanisms, we get an inkling of the complex functional system that must be created and organized out of highly specialized proteins in order to maintain the life of the cell.

As mentioned earlier, cells are made up of 70 percent water and 26 percent large molecules such as proteins. In fact, the proteins constitute half the weight of the dried material. It is clear that for the growth, maintenance, and development of the cell structure, a process of building up proteins must take place.

The transportation mechanism from outside the cell provides the construction materials, such as cholesterol, amino acids, and other nutrients. But the assembly of those building materials takes place inside the cell, in the protein factories called ribosomes, and is intimately connected to the cell's own DNA genetic code instructions.

The actual manufacture of proteins inside the cell requires the presence of a host of specialized protein molecules called enzymes. The specific function of an enzyme is to speed up certain chemical reactions by a factor of 1,000 or more. The enzymes perform this task by assuming specific contorted geometric shapes to provide "meeting places" for the participants in certain chemical reactions. Instead of relying on a chance encounter in a solution where the chemicals are usually spread far apart, the enzyme attracts the participants to specific locations, and thus the reaction can take place much faster. There exist about 2,000 different enzymes in the various life forms on Earth today.

One such special enzyme is RNA polymerase, a very complex enzyme made of about 5,000 amino acids. Its function in the mechanism of protein synthesis is to attach itself to that place on the DNA double helix where the instruction for building a specific protein begins. Then an astoundingly complex process begins: the DNA transcription process, which we shall sketch only in broad outline. The RNA polymerase begins to unwind the two strands of the DNA helix and to build an RNA duplicate of that specific DNA segment containing the genetic instructions for building the particular protein.

Certain chemical reactions are initiated to start the process when the RNA polymerase attaches itself to what is known as the promoter site. Similarly, certain chemical reactions stop the process when the RNA polymerase—as it moves along each rung of the DNA strand—reaches a termination signal. The end result is an RNA copy of the DNA protein assembly instructions. This copy is called a messenger RNA (mRNA), which then floats in the cell fluid toward an empty ribosome. The freed RNA polymerase may then go on to build other mRNA molecules or even ribosomes or something called transfer RNA (tRNA). The whole process is quite fast—about thirty genetic letters can be strung in the mRNA in one second at 37 degrees centigrade.

The messenger RNA attaches itself to an empty ribosome so that the actual protein assembly process can begin. Surrounding the ribosome, with its now-attached mRNA, is a swarm of amino acids attached to transfer RNAs. Each of the twenty different amino acids in the cell fluid can attach only to a particular type of transfer RNA, which contains the three genetic code letters corresponding to the particular amino acid in question.

Another complex chemical process begins, for which some of the physical-chemical details are still unclear to the molecular biologist. The gist of this process of actually manufacturing the protein is as follows. (See Figure 3.1.)

The ribosome unit clamps the messenger RNA at a beginning site to initiate the process. The first amino acid link in the protein chain is delivered by a transfer RNA that carries the right genetic letters to match the first three instruction letters in the messenger RNA.

Once that transfer RNA has delivered its amino acid correctly in the protein sequence to be built (say, Phe in Figure 3.1) the ribosome will move its action on to the next three letters in the messenger RNA instructions. Once there, it will attract the corresponding transfer RNA and its amino acid (say, Trp in Figure 3.1). This amino acid hooks onto Phe, and the emptied Phe tRNA is free to leave.

Just as the RNA polymerase moved along the original DNA strand to create the messenger RNA, so the ribosome moves along the messenger RNA to create the protein. In a bacterium the synthesis of a protein of 400 amino acids is completed in about twenty seconds (Alberts et al. 1989, p. 212). When the protein chain is released, the chain will coil and fold. This vastly compacted form of the protein specifies the protein's function.

This process for manufacturing proteins is economical, elegant, and ingenious. However, the complexities of the organized chemical activities are staggering. The unraveling of the essential details of this process stands as

PROTEIN
SYNTHESIS

Figure 3.1

Protein synthesis starts as an RNA strand in the cell unwinds the specific place in the DNA helix where the instructions are found for making a particular protein. A copy of the instructions is generated on an RNA snippet (messenger RNA, or mRNA). This mRNA drifts through the cell interior until it attaches to the cell's manufacturing plant—the ribosome enzyme. The end of the mRNA snippet is fed into the ribosome and acts as a kind of "tape" from which the codon sequence is read by the ribosome.

Surrounding the ribosome are other RNA snippets (transfer RNA, or tRNA) that carry the various amino acids needed to build that protein. One cycle of the manufacturing steps is outlined in the adjacent figures.

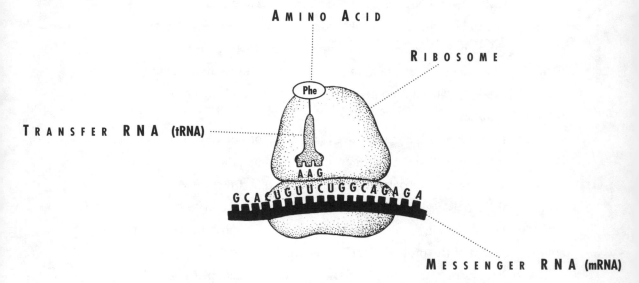

Initial configuration of ribosome enclosing the messenger RNA strand and a
transfer RNA with its attached amino acid (phenylalanine).

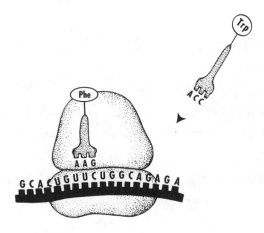

STEP ONE

A transfer RNA (tRNA) approaches the ribosome enclosing the messenger RNA (mRNA) strand.

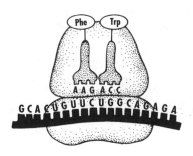

STEP TWO

The tRNA with its attached amino acid Trp (tryptophan) binds with its ACC site to the corresponding UGG site on the mRNA.

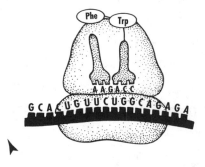

STEP THREE

The amino acid Trp links to the previous amino acid Phe (phenylalanine), which is already attached to the protein chain being built.

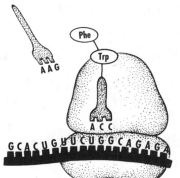

STEP FOUR

The ribosome moves three steps to the right on the messenger RNA strand, ejecting the empty transfer RNA, which carried the amino acid Phe. The system is now ready for the next cycle of amino acid attachment to the protein chain.

one of the great intellectual achievements of this century. And yet, some of the details are still to be understood.

The consensus among molecular biologists (see introductory chapter, Alberts et al. 1989) is that the formation of the single-stranded RNA preceded that of DNA in the primordial broth; this well may be the case. Somehow, then, the double-stranded DNA was formed out of the single-stranded RNA.

In Chapter 2 we saw that the molecular structure of the RNA and DNA strands was quite monotonous: one sugar-phosphate molecule attached to another, much like a string of beads, with connecting rungs between two identical strings. These rungs are made from the four simple nucleotide molecules for the DNA strands (A, T, G, C) and four for the RNA strands (A, U, G, C).

In contrast, the proteins are made up of twenty different amino acids in a large number of combinations that allow for a dazzling variety of characteristics. There are 200,000 different proteins used in the life forms we know today. Each protein has to have its set of assembly instructions coded in a DNA strand. Each protein has been invented *for a specific purpose.*

If we look at the elaborate complexity of proteins like RNA polymerase—with its 5,000 amino acids, many of which went into the first cells created 3.8 billion years ago—a question inevitably arises: Was there really time enough for Darwinian chance and natural selection to create these complex molecules in the course of only 400 million years?

The question is further aggravated by consideration of the cell itself. How does chance build up a cell—any cell? To achieve that feat, phospholipids—masses of them—first had to be created in the primordial broth. Then they had to meet in large numbers to form the cell membrane. Then the cell membrane had to "accidentally" enclose RNA or DNA. Furthermore, this DNA had to learn by chance to program a variety of types of proteins, some of which had to be inserted by chance into the cell membrane to act as globular or helix protein receptors and transmitters. Still others had to be created by chance to form the protein factories in the cell interiors. Finally, the DNA had to learn to replicate and build a new cell by using RNA polymers and other forms of RNA. All this had to be achieved in the relatively short span of 400 million years—and we haven't even touched on the complicated chains of chemical reactions that go into a cell's metabolism, respiration, or photosynthesis.

These processes, if encountered in any human manufacturing plant, would be considered remarkably high-tech. The Darwinists and neo-Darwinists have never been able to use hard mathematical calculations based on probability theory to uphold the assertion that chance could realistically be the creative agent. The words of highly respected biologist George Wald, "Given so much time the 'impossible' becomes possible, the possible probable and the probable virtually certain," simply do not suffice anymore (Wald quoted in L. Cudmore 1977, p. 138.).

Even in the monumental *Molecular Biology of the Cell* (Alberts et al. 1989), we read:

> The molecular process underlying protein synthesis seems inexplicably complex. . . . The complexity of a process with so many different interacting components has made many biologists despair of ever understanding the pathway by which protein synthesis has evolved. However, the recent discovery of RNA molecules that act as enzymes has provided a new way of viewing that pathway. . . . Over the course of evolution, individual proteins could have been added to this RNA machinery, each one making the process a little more accurate and efficient.

We see here, even in a treatise on molecular biology, the classical Darwinian conjectural "handwaving" without a shred of mathematical proof.

In Chapter 2 we learned that building the DNA instructions for a simple protein made up of only thirty-four amino acids through chance encounters in a primordial broth would in all probability take far more time than the age of the Universe. Thus, to arrive at the mechanism of protein synthesis by chance would seem to be out of the question.

Furthermore, these mathematical-physical exercises touch upon only the issue of attaching a certain number of genetic letters (A, T, C, G) into a specific sequence in the DNA molecular strand. The overriding higher-level issue is how to incorporate *idea structures* into these DNA sequences.

If these genetic code instructions came about by chance, they are trivial. But if humans, in deciphering the genetic code, have in fact discovered "God's Own Language," these instructions are awesome.

In a strange way, by creating an oxygenated atmosphere, the prokaryotic cells prepared themselves for their own evolutionary stagnation. Only vestiges of the anaerobic prokaryotes still exist today in oceanic floor vents or volcanic fumaroles on land. The prokaryotic cells did continue to evolve to a limited extent by developing oxygen-resistant strains, such as aerobic bacteria. These abound on Earth today, and they form vital parts of any ecosystem, but they never advanced to form multicellular structures.

The oxygenated atmosphere opened the evolutionary pathway for the development of the eukaryotic cells having even more complex cell structures, a necessary condition for the creation of multicellular structures like ourselves. The human race sits at the top of an immense evolutionary pyramid, at the base of which lies the "humble" prokaryotic cell that made it possible for oxygen-breathing life to exist on this planet in the first place.

The study of microfossils in the 1950s and the 1960s demonstrated that around 1.4 billion years ago, the primitive prokaryotic cells (those without a central nucleus) were gradually evolutionarily surpassed by the eukaryotic cells (those with a nucleus). Undoubtedly, millions of years were needed to prepare for this development; so far no earlier fossil traces of eukaryotic cells have been found.

We do not have a clear picture of the internal structure of the first eukaryotic cells, but even the rudimentary internal structures found in fossils point to a definite kinship with many of the structures observed in today's eukaryotic cells. And although such cells come today in many varieties, they all share a basic internal functional pattern that may best be reflected in he internal structures of the still-living one-celled eukaryotic life forms such as amoebas and other protozoans.

To accommodate a more complex internal structure, the eukaryotic cell has to be considerably larger than its forerunner, the prokaryotic cell. The eukaryotic cell is typically 10 to 100 times larger than the prokaryotic cell and correspondingly contains a much larger number of atoms. The average prokaryotic cell may have on the order of 100 million atoms; the eukaryotic cell contains on average 1 million times more atoms, or 100 trillion atoms. *Thus, there are 1,000 times more atoms in a single eukaryotic cell than there are stars in the Milky Way.*

The common "blob of protoplasm," the amoeba, is a typical representative of the current descendants of the first eukaryotic single cell. It is so small—only one-tenth of a millimeter across—that it is barely visible as a speck to the unaided eye and is comfortably visible only under the microscope. Yet it is an agglomeration of 100 trillion atoms in a highly ordered structure. It has a "skin," the cell membrane, maybe one hundredth-thousandth of a millimeter thick.

Within its cell membrane the 100 trillion atoms are built into a veritable molecular community of great diversity and complexity. That single-celled amoeba lives as a totally self-contained system. Whereas we human beings depend on all our 100 trillion cells (each having 100 trillion atoms) to function harmoniously together, the amoeba tranquilly rests within itself in its own single cell.

In the prokaryotic cell structure, we saw the rudiments of an *idea structure* being imposed on the "dead" matter of atoms and molecules: the *idea* of transport mechanisms in and out of the cell membrane, the *idea* of digesting nutrients and excreting waste material, the *idea* of manufacturing protein, even a rudimentary *idea* of locomotion.

This idea structure reaches a remarkable flowering in the material manifestations of the amoeba's internal structure, as well as those in modern animal and plant cells (see Figure 3.2). The "old" idea structures in the

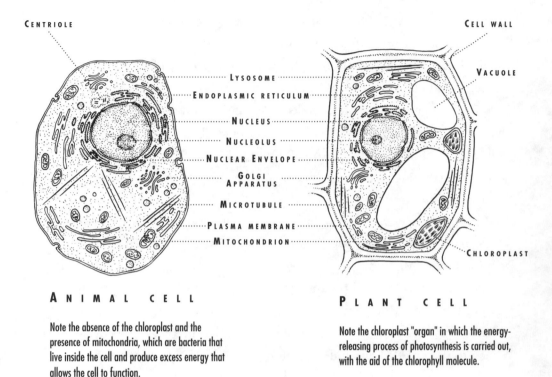

CENTRIOLE

CELL WALL

VACUOLE

LYSOSOME

ENDOPLASMIC RETICULUM

NUCLEUS

NUCLEOLUS

NUCLEAR ENVELOPE

GOLGI APPARATUS

MICROTUBULE

PLASMA MEMBRANE

MITOCHONDRION

CHLOROPLAST

A N I M A L   C E L L

Note the absence of the chloroplast and the presence of mitochondria, which are bacteria that live inside the cell and produce excess energy that allows the cell to function.

P L A N T   C E L L

Note the chloroplast "organ" in which the energy-releasing process of photosynthesis is carried out, with the aid of the chlorophyll molecule.

PLANT AND ANIMAL CELLS

Figure 3.2

A major watershed in evolution took place 1.5 billion years ago with the appearance of the eukaryotic cells containing a nucleus that encapsulated DNA strands. Furthermore, these cells opted for two radically different evolutionary pathways: Some cells utilized photosynthesis for their energy needs (future plants), and other cells ate other life forms for their energy needs (future animals).

prokaryotic cells have been translated into more elegant, elaborate, and functionally efficient physical manifestations.

Within these cells resides a magnificent array of organelles, small "organs" that would be the equivalent of the heart, the digestive tract, the brain, and so on in a multicellular animal. There exist about a dozen organelles in a fully developed eukaryotic cell.

In the center of a cell resides the nucleus, which contains the DNA double helix and the nucleolus, surrounded by a protective second membrane called the nuclear envelope. Outside this nucleus reside a variety of other organelles like the mitochondria, which power the cell. There are also thousands of ribosomes, which are the protein "factories"; the lysosomes, which represent the "digestive tracts"; and the smooth and rough endoplasmic reticulum, the microtubules, and their aggregates, the centrioles, which form the interior "skeletal" support structure. The microtubules may pierce the cell membrane to form rudimentary "legs"—the flagella—which provide locomotion. Finally, there are the Golgi apparatuses, which serve as "warehouses" and "dispatchers" for the proteins created by the ribosomes.

Whereas the proteins and enzymes for prokaryotic cells were aggregates of several thousand atoms, the cell organelles are aggregates of several million atoms. Correspondingly, the genetic code instructions for the cell to manufacture those organelles must be extraordinarily complex to enable the building of those intricate geometrical structures and functions.

We have not yet touched upon the details of locomotion. In both the prokaryotic and the eukaryotic cells, locomotion stems from the development of the threadlike flagella that protrude outward from the cell membrane into the surrounding watery environment. These flagella are rudimentary limbs by which single cells may propel themselves through liquids.

In conjunction with an article on the human engineering field of nanotechnology (exceedingly small structures), the November 1992 issue of *Scientific American* touted a photo of the world's smallest electromotor fitting inside the eye of a needle. As remarkable an achievement as this may be, Nature's flagellum rotor engine is 10,000 times smaller.

The propelling mechanism of a flagellum is so remarkably ingenious that even a sober graduate school treatise like *Molecular Biology of the Cell* includes the following expression of amazement:

> Bacteria swim by means of flagella that are much simpler than the flagella of eucaryotic cells . . . and consist of a helical tube containing a single type of protein subunit called *flagellin*. Each flagellum is attached at its base, by a short flexible hook, to a small protein disk embedded in the bacterial membrane. Incredible though it may seem this disc is part of a tiny "motor" that uses the energy stored in the transmembrane H^+ gradient to rotate rapidly [100 revolutions per second] and turn the helical flagellum. (P. 720)

The cilia and flagella have amazing geometrical intricacies built at the atomic-molecular level to enable those organs to form and to define their functions. What prompts the cell to build those tubular structures, aggregate them in centrioles, and then attach them to the cell membrane to grow outward as limbs? It is as though an intelligence first conceived of the *idea* of locomotion and then translated it into a viable physical solution.

The last organelle we will discuss is the cell's "power plant." This organelle represents another crucial development in the evolution of life. We have already mentioned that there exist in Nature two distinctly different metabolic pathways: one is aerobic and uses oxygen, and the other is photosynthetic and uses carbon dioxide, water, chlorophyll, and light.

In the development of the eukaryotic cells 1.5 billion or more years ago, some eukaryotic cells incorporated an aerobic bacterium, the mitochondrion, into their internal cell structure, whereas other eukaryotic cells incorporated a green algae having chlorophyll. Those cells with the mitochondria were the ancestors of all animals, and those with the green algae were the ancestors of all plants. That vital development and evolutionary decision some 1.5 billion years ago set the stage for the bifurcation of life as we know it today: a world of plants and of animals that need the plants for oxygen and sometimes food. The aerobic bacteria survived intact inside the eukaryotic cells as mitochondria, whereas the green algae developed into the organelle known as the chloroplast.

The mitochondria are almost as large as the nucleus of the cell, which makes them among the largest organelles. Their number varies and is as high as 1,000 in especially strong energy-producing cells. Mitochondria were first observed in 1850, but their function was a mystery until the 1950s, when it was realized that the mitochondria provide energy for the cell. Carbohydrates, fats, protein molecules are broken down by enzymes in the cell interior, providing fuel for the mitochondria.

A eukaryotic cell with internal mitochondria is an extraordinary example of a symbiosis of two single-celled life forms that has persisted for 1.5 billion years. We have about 100 trillion of those aerobic bacteria living inside us, powering almost all the cells in our bodies.

The internal structure of the eukaryotic mitochondria is very different from the structureless gel of the ordinary free-living aerobic bacteria. The mitochondria have been streamlined to become exceptionally efficient energy producers: they provide more energy than they need. The mitochondria also appear to regulate the concentration of water, calcium, and other ions in the liquid gel of the host cell.

The cellular ancestors of animals and plants we have just described flourished for hundreds of millions of years as single-celled life forms in the oceans. During that time, thousands of varieties of single-celled life forms developed. Biologists call those life forms protists or protozoans.[2] Today, more than 10,000 species are known.

The life forms grouped themselves into two camps. The eukaryotic cells with the chloroplasts could survive by themselves and so are called autotrophs. They were "self-feeders" that nourished themselves on inorganic materials and derived their energy from sunlight. The other camp included the

[2]Margulis and Schwartz (1982) prefer to combine them into one kingdom—the protoctista (p. 77).

heterotrophs, which fed on the autotrophs and derived their energy from decomposing the carbohydrates of autotrophs by breathing oxygen. Thus, the two major kingdoms of life (ancestral "plants" and "animals") were created over a billion years ago. By that time, 80 percent of the lifetime of Earth had passed—the long preparatory steps had finally been taken for an eventual "assault" by multicellular animals on the barren landmasses.

One excruciatingly important evolutionary step remained to be taken: the invention of sex. This development took place relatively early in the unfolding development of the eukaryotic cells, maybe about 400 million years after the first such cells were formed, or 1.1 billion years ago. The exact date is, as always, difficult to pinpoint And how sex came to be is also still a mystery. Even today, not all of the exact molecular processes in sexual reproduction have been charted.

The invention of sex implied that two different single cells would "pool" their DNA codes. Out of that pooling, or mating, interesting new molecular combinations could be formed when genetic messages were exchanged, combined, or modified. The sexual process sped up the evolutionary processes, although, as we shall see in later chapters, it does not get the chance hypothesis off the hook.

Some of the eukaryotic species have continued to exist to this day as the fascinating single-celled protozoans. They all, by necessity, live in water. Many of us remember biology lessons in school in which we examined those life forms in a drop of water culled from a nearby pond. When seen through the eyepiece of a microscope, that drop of water teemed with protozoan life, from gentle amoebas to darting paramecia to "carnivorous" stentors. Books have been written about the marvelous variety of figures, forms, locomotion, and behavior of those delightful little single-celled creatures. And many a biologist has been enamored with them, like Larison Cudmore, who wrote this exuberant description in her book *The Centre of Life:*

> Amoebas may not have backbones, brains, automobiles, plastic, television, Valium or any other blessings of a technologically ad-

vanced civilization; but their architecture is two billion years ahead of its time. The amoebas had the architectural ideas of R. Buckminster Fuller before there was anyone around capable of having an idea. The amoebas are the Bauhaus, the Taliesin of the West of the protist world; though apparently Frank Lloyd Wright never talked much about them. . . . And as usual in any art there has to be an over-achiever. Il sorpasso. Our amoebic Beethoven, our gelatinous Gaudi is Hexacontium asteracanthion. One geodesic dome will not do for this superarchitect; it has to be three lacelike fretted glass domes, one inside the other, like the follies that are the pride of oriental ivory carvers; detailed hollow spheres nested within each other, the openwork of one revealing several more exquisitely carved balls. (P. 15)

Most of us would agree with Cudmore that the world of those protozoans is one of incomparable beauty. (See Figure 3.3, which shows the architecture of *Hexacontium asteracanthion* as opposed to a photo of the vaulted ceiling of Westminster Abbey in London.) Those single-celled life forms have developed an ability to construct the most exquisitely shaped "houses of glass" for their protection. The blind god of chance surely has no need for such beauty, since only the quality of function matters in the battle for survival of the fittest.

The physicists often precede the biologists in a search for understanding the deeper implications of simplicity, elegance, and beauty in the mathematics that seemingly govern the behavior of inanimate matter. One of the founders of quantum mechanics, Nobel Prize winner Werner Heisenberg, delivered an address in 1970 on "The Meaning of Beauty in the Exact Sciences" to the Bavarian Academy of Fine Arts. At the end he paraphrased Johannes Kepler to represent the concept of beauty in the sciences: "Mathematics is the archetype of the beauty of the world." Interestingly enough, to represent beauty in the fine arts he chose a quotation from the philosopher Plotinus in the third century A.D.: "Beauty is the translucence, through the

The POETRY *of*
BIOLOGICAL DESIGN

Figure 3.3

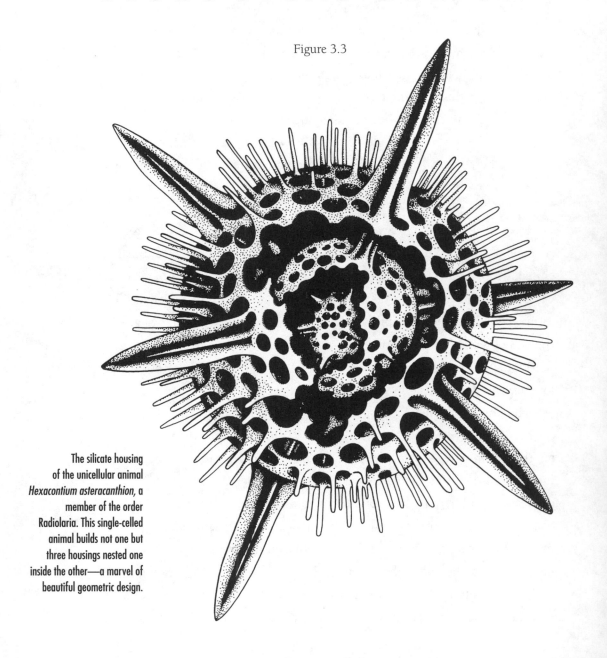

The silicate housing of the unicellular animal *Hexacontium asteracanthion*, a member of the order Radiolaria. This single-celled animal builds not one but three housings nested one inside the other—a marvel of beautiful geometric design.

THE UNICELLULAR ANIMALS
(protozoa)

These single-celled animals have developed a large variety of forms over the past 1.5 billion years. About 25,000 species are known today; 7,000 are parasites in humans and animals. Many protozoans form a vital part of ordinary soil. One gram of earth (about one-tenth of a teaspoonful) may contain 100,000 flagellates (such as *Giardia*), 50,000 amoebas, and 1,000 ciliates (such as *Paramecia*).

One particular order is composed of the radiolarians, with their silicate housings. They occur in vast numbers in the oceans, and over 5,000 living species have been classified. Their dead skeletons sink to the bottom of the sea to form "radiolarian ooze." One particular bel of this ooze in the Pacific Ocean covers an area of about 750,000 square kilometers—the size of Germany and Italy put together.

The fan vaulted ceiling of Henry VII's chapel in Westminster Abbey, London. This early sixteenth-century type of architecture is a phase of late Gothic style, which is unique to England.

material phenomena, of the eternal splendor of the 'One.'" As we continue to peer through our windows of knowledge, perhaps we shall see these two concepts of beauty and function fuse in the emergence of a new explanatory paradigm for evolutionary biology.

The CAMBRIAN FLOWERING

During little more than 15 percent of life's span on Earth, all the animal and plant forms we know today were developed. The amount and variety are staggering. Biologists estimate that today more than 3 million different species are living on Earth, many still unnamed and undiscovered. If we reckon—as some biologists do—that each species has on average a lifetime of 10 million years, then more than 200 million species must have been created over a period of only 700 million years (the period during which fossils of multicellular life forms have been found). Richard Lewontin of Harvard University, however, estimates that 2 billion species have appeared on our planet since the beginning of multicellular life forms (*The Fossil Record and Evolution*, p. 17).

It took about 1 billion years to create one "species," the prokaryotic single cell, from the primordial soup. It took another 2 billion years to create yet another single new life form, the eukaryotic cell, together with assorted energy cycles. And then, about 700 million years ago, a little before the Cambrian geological era started, heaven burst open, and all kinds of weird, fantastically shaped multicellular life forms poured onto this planet. Whoever dreams up extraterrestrials and science fiction monsters has not a wit of imagination compared with this enormous profusion of life forms—more than 200 million of them, each species built up of millions of cells, sometimes trillions of cells, and each cell made up of 100 trillion atoms! Is it really meaningful to state, as Darwinians do, that such a boundless imagination can be displayed by chance? An impression of the dazzling increase in the speed of evolution can be gleaned from the fact that the time from the appearance of Lucy, the first primitive hominid, to the genius of Leonardo da Vinci was a mere 4 million years. This is less than one-thousandth the age of our planet. For 99.9 percent of the time that life has existed on Earth, there were no Lucys or Leonardos.

Our next window of knowledge will show us certain aspects of that Cambrian explosion of life forms. It is impossible to do justice to the large body of knowledge about ancient life forms culled from the study of fossil records. All that we know today of life in the history of our planet has really been accumulated only during the last 100 years. Paleontology is truly one of the youngest of all sciences. Nevertheless, we will attempt to paint with broad strokes an outline of the development—the evolution—of the life forms, from the first appearance of primitive cellular groupings to the superbly streamlined evolutionary forms of modern mammals.

In the history of single-celled life forms there were several evolutionary breakthroughs: the prokaryotic cell, metabolism, anaerobic photosynthesis, aerobic photosynthesis, the eukaryotic cell, the mechanism of sex. Those were also basic tools for forging the creation and evolution of multicellular life forms, but there remained many tools to be invented, including the sticky proteins that could cement cells together into multicellular units.

For the animal forms, the protein collagen was produced. It is a long, fibrous protein that gives rigidity and firmness to muscles and tendons. Another protein, elastin, is elastic in nature and forms part of the skin and blood vessels. Reticulin is closely related to collagen and helps bind the cells together and give organs their shape.

In the plants, a different solution to the structural problems of joining single cells into multicellular structures was found. The plant cells were surrounded by a primary wall and a secondary wall. The primary wall of one cell is then joined to the primary wall of another through a middle lamella, a kind of intercellular cement that contains pectin. (Pectin is what is used to make jams gel.) The primary wall forms the outer covering of the cells of fruits, roots, and fleshy stems. The inner, secondary wall gives the plant cells strength and mechanical support. The secondary walls of plant cells provide humans with, for instance, lumber for building and cotton for clothing.

If we go back over the history of evolutionary breakthroughs, we might find it reasonable that the invention of the first cell membrane out of the primordial soup could have occurred by chance. Primitive and current cell membranes are based on the natural tendency of lipid molecules (like those in oil) to produce a "corral": the molecules' water-repelling heads point outward, fending off the external water molecules.

However, not even the entire age of the Universe would be enough time for chance to produce the diversity of complex organic molecules and the resulting complex life forms. It is fine for the Darwinian biologist to state that mutations induced genetic changes to produce new materials that were tested by the natural selection of the outer environment. But the chance operator is a blindfolded manipulator of a vastly more complex cube than Rubik's, and a few hundred million years simply will not suffice for those intracellular glues and reinforcing structures to be produced by chance.

Even the simplest of multicellular forms are made up of several different kinds of cells, whereas the most complex ones may harbor as many as 200 different types of cells. Figure 4.1 depicts some of the differently shaped cell forms and types that go into making the human body: the egg cell, sperm cell (with its flagellalike tail—a relic from primeval days?), hair cell, eye rod cell, bone cell, and muscle cell. The variety is mind-boggling. Each cell form is honed to serve a specific function.

Such wide cell differentiation had to be created during only a few hundred million years. Each cell is made up of trillions of atoms, which have to be grouped into immensely complex and exacting geometric structures; it must be made precise not only in structure but also in function. Then those

SINGLE CELL TYPES

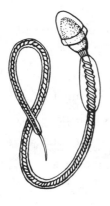

S P E R M   C E L L

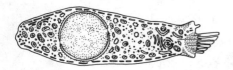

H A I R   C E L L

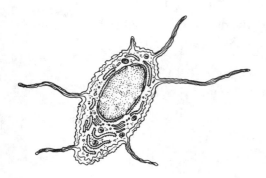

B O N E   C E L L

In multicellular organisms, the simple oval forms of single eukaryotic cell membranes have been distorted into a variety of shapes. The human body contains over 200 different types of cell forms. This figure depicts (not to scale) six of these human cell types: sperm cell, hair cell, bone cell, muscle cell, egg cell, and rod cell in the eye.

As a result of their functional specialization, the sizes of these cells vary greatly from the pinpoint size of the sperm cell to the 3-foot length of a nerve cell (see Figure 6.1,

Figure 4.1

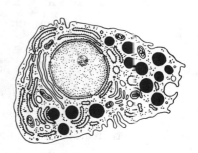

M U S C L E   C E L L

page 124) that connects the spine and a toe.

The organic materials found in the membranes

vary depending on the cell type—from bone

protein to the materials of hair or muscle cells

to the pliable liquid molecules of sperm and

egg cells. In addition, for the different cell

types, a large number of specialized molecules

have evolved to enable the cells to communi-

cate with each other as needed for the whole

organism to function. The human body contains

approximately 100 trillion cells—all orches-

trated into one magnificent whole.

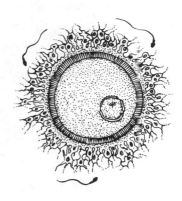

E G G   C E L L

R O D   C E L L   I N   E Y E

cells must be made to communicate and interact so that the whole aggregate of cells functions as one integral unit.

Today, we see in the species called *Volvox* a living relic of how multicellular life forms may have begun more than 700 million years ago. A *Volvox* colony is a hollow sphere about the size of a pinhead, made up of a large number of single cells arranged so that their individual flagella lie on the outer surface of the sphere. The joint coordinated action of those flagella propels the sphere in whatever direction the cell group's "consciousness" dictates. The cells that make up the hollow sphere are no different from single-celled creatures that lead individual, separate existences. Hence, there is little or no cell differentiation in the *Volvox*.

However, the principle of cell *coordination* may have started in that way and developed further in the sponges that are believed to have appeared some 800 million to 1 billion years ago. In the sponge, the flagella are used not for locomotion but to whip water through surface pores into the sponge. Inside the sponge, cells extract nutrient particles from the water and then expel the wastes through larger vents.

Some sponges produce a special substance around their cells, providing the beginnings of a multicellular glue that gives a supportive structure to the organism. That glue enables a common bathtub sponge to retain its structure when the resident cells of the sponge have been boiled and killed.

Other sponges, like the single-celled and multicellular radiolarians, create exquisite glass needle structures for support. In the words of biologist David Attenborough, in his book (p. 29) based on the television series *Life on Earth:*

> When you look at a complex sponge skeleton such as that made of silica spicules, which is known as the Venus Flower Basket, the imagination is baffled. How could quasi-independent microscopic cells collaborate and secrete a million glassy splinters and construct such an intricate and beautiful lattice? We do not know.

Perhaps the time has come for biologists to expand their mental universe and admit into their ken the possibility of a consciousness existing within the groups of cells as a whole, even in such a "lowly" creature as the Venus Flower Basket. Why not embrace the hypothesis that this unfolding of creative inventiveness is indicative of an *overarching* intelligence, "playing" thoughtfully, slowly maturing in concepts and solutions as it moves from discovery to discovery. This hypothesis inspires much more awe and excitement than the mere stumbling progress of chance.

Although the sponges evince use of cellular glues to form collective cell structures, they do not display cell differentiation into such types as specific muscle cells or nerve cells. The first life form to make use of differentiated cells—cells that are unable to live by themselves—appears to have been the jellyfish. Up until the 1940s, paleontologists considered a puzzling abrupt cutoff in fossil records at the beginning of the Cambrian era some 560 million years ago. This cutoff is observed everywhere on Earth and must mean that the hard-bodied life forms appeared almost simultaneously all over the planet. Why does this cutoff exist?

If soft-bodied forms preceded the hard-bodied ones, their remains would tend to be obliterated and normally would not be fossilized. However, the discovery of fossils of soft-bodied life forms—including the jellyfish—in 1947 by Australian geologist R. C. Spriggs in southern Australia represented a rare exception. Tests of those geological formations place the deposition of those fossils at around 650 million years ago, well before the Cambrian era began. Similar fossil depositions were later found in South Africa and England.

A jellyfish today shows quite a complex multicellular structure, with none of the individual cells able to live by themselves. Basically, its structure is formed by two layers of cells separated by a jellylike glue that gives a certain rigidity in water. Some cells are modified to make contractive motions thus representing a rudimentary form of muscle cells. Others have been modified to transmit electric impulses in a network of cells, resembling a rudimentary form of a nervous system. Finally, a third group of cells have developed into an attack or defense system: the stinging cells release threads that end in a miniature harpoon structure and are sometimes loaded with poison.

Already at that primitive level, quite a remarkable diversification of cell forms and functions had developed. Even though we base our observations

on the characteristics of today's species, it is very likely that similar characteristics appeared in the ancient ancestors of the jellyfish, since their fossil
shape is unmistakably that of a jellyfish.

Even more remarkable is the reproductive cycle of the jellyfish. When a
jellyfish egg and sperm meet in the sea, the fertilized egg does not produce
another jellyfish. Instead, the fertilized egg settles down at the bottom of the
shallow coastline and develops into a tiny flowerlike organism, called a
polyp, which is an intermediary life form. It is not free-swimming at all but
instead is attached to the ocean floor, where it feeds itself with tiny flapping
cilia. After a while, the polyp will bud, and the bud produces a small jellyfish that will then take up a free-swimming lifestyle.

If sex was "invented" some 900 million to 1 billion years ago and the
eukaryotic cells 1.5 billion years ago, the rapidity with which the jellyfish's cell differentiation and complex reproductive cycle must have developed is utterly amazing—requiring only 200 to 300 million years. This
evolutionary development is even more remarkable if we consider it at the
molecular level: the new proteins to be developed, the new cell types to
be made, and the whole idea of an intermediary stage. How could all that
be realized by chance mutations in the nuclear genetic code, the DNA, in
such a short time? No mathematical answers have been given by the
Darwinists.

Biologists divide all life on Earth today into five kingdoms. The Monera
are the prokaryotic single-celled forms such as bacteria and blue-green
algae. The Protistas are the single-celled eukaryotic forms such as amoebas
and protozoans. The Fungi are intermediate between the animals—the Metazoans—and the plants—the Metaphyta—because Fungi ingest organic materials. The Metazoans are divided into twenty-six categories termed *phylae*.
At least another nine phylae have become extinct. Of the twenty-six living
phylae, fourteen represent worms in one shape or another. It is obvious that

a lot of "thinking" and experimentation have gone into the creation of worm-like life forms.

In our brief, panoramic exposé of the development of multicellular Lfe, it must suffice to give a fleeting impression of the beginning of multicellular life forms. The development of such forms centered for a few hundred mil-lion years on the flatworms, which represent one phylum in the animal kingdom. In Chapter 3, we learned about the manifestation in single cells of complex idea structures like eating, digestion, and waste disposal, to which we could add reproduction. We shall see that these idea structures likewise condition the design of multicellular life forms such as the flat-worms.

The flatworms are very primitive forms indeed. Like the jellyfish, they have only one opening to their digestive tract that serves both to take in food and to eject waste. However, the flatworms have three tissue layers in-stead of the two in the jellyfish. The flatworms have an outer protective tis-sue layer, an inner layer lining the gut, and an intermediary layer forming the core of a muscular "body" surrounding the gut. They breathe through their skin; gills had not been invented when flatworms first appeared. And their limbs of locomotion are cilia, hairs that cover their underside so that they can glide over surfaces or burrow in the sand. We see how Nature makes extensive use of earlier inventions: The cilia were first developed for locomotion in the protozoans.

At the front end of the flatworm is an opening that receives nutrients found in water. That spot can be said to be the beginning of a mouth and a head. Also in the front end are a few light-sensitive spots, the beginning of eyes. These rudimentary eyespots are linked to muscles by a primitive net-work of nerve fibers so that the animal can react to what it sees. In some of the nerve fibers there are thickenings—the beginning of a brain.

It is fascinating to look at that design as a very crude sketch of an ani-mal, such that the basic functions are laid out but the final form has yet to be finished. The basic ideas of a gut, a body with limbs for locomotion, eyes, and a brain all appear simultaneously in primitive form. A rough conception of an animal has already materialized in this life form, again prompting us to wonder how it is possible for chance to orchestrate such a design.

The flatworms represent the beginning of the invertebrate animals, those without skeletal structure. Today there are more than 3,000 species of flatworms, not to mention the numerous species of the thirteen other phylae of worms. Evidence suggests that the flatworms first appeared between 600 million and 1 billion years ago (see David Attenborough, *Life on Earth,* p. 39).

The Burgess Shale, a group of fossil findings from British Columbia in western Canada, demonstrates the variety of invertebrate forms having evolved up to the Cambrian era. Fossils of more than 120 marine invertebrates were found. (See the book by Stephen Jay Gould, *Wonderful Life: The Burgess Shale and the Nature of History.*) Those fossils were laid down some 550 million years ago.

Some of the worms developed protective shells, just like some of the protozoans. Those worms were the ancestors of the extensive phylum of molluscs, which today comprises some 60,000 species. Other worms developed segmented bodies; fossils of the early primitive segmented worms are found in the Ediacaran sediments deposited in southern Australia 680 million years ago. Segmentation represents a remarkable advance in the design of animal bodies. The Ediacaran fossils appear amazingly similar to the bristleworm living in coral reefs today. That creature is made up of a chain of "modular" segments. Each segment is a self-contained unit with a pair of legs on which hairy appendages for respiration are found. Within the body walls are a pair of tubes to secrete waste. Through the whole chain of segments, there runs a common digestive tract, a large blood vessel, and a nerve chord, linking the segments and coordinating them. At the front is a real head.

Such changes were a vast improvement on the first crude animal design that appeared in the flatworm, and they were precursors of the further developments to come. In the Burgess Shale, besides fossils of segmented worms, there are also fossils of trilobites, an enormously successful species that survives today in the form of the horseshoe crab. The trilobites flourished during a period of 300 million years, from roughly 550 million years ago until about 250 million years ago. They evidently represented a very successful design because they survived twice as long as the dinosaurs In the design of the trilobites, there was a very important chemical invention: the development of the substance chitin. That material was later used in the construction of insects, by far the largest group of life forms existing today. Some 700,000 species of insects have been classified, and probably two or three times that number still await classification.

The insects are also descendants of the segmented worms. About 400 million years ago, through a series of astounding inventions such as air breathing and chitin-encapsulated bodies that could retain water, insects moved onto land together with the first plants. The first land-crawling in-sects were millipedes, and some of them grew as long as a cow, or 7 feet in length.

Another major group of life forms found today—the crustaceans, such as shrimp, crabs, and lobsters—also evolved from the segmented worms. However, they developed not one pair of antennas, as did the trilobites, but two pairs. Today, there are some 35,000 species of crustaceans—four times the number of bird species.

Not much is known from fossil records of the early development of the plant kingdom—the Metaphyta—in their marine environment. Undoubt-edly, a series of experiments were made of which we have no record. The most important invention in this kingdom was the development of a waxy covering, the cuticle, on the outside surfaces of plants. The cuticle prevented water from evaporating from the plant cell interiors. The plants also had to solve the problem of reproduction, which they did by adopting asexual d -vision and, when water was available, sexual reproduction.

The first plants on land were akin to the now-living mosses, liverworts, and some leafless branching strands. The strands adopted another very im-portant invention: thick-walled cells. This development made it possible for water to be pumped up in the plants' stems and thus enabled the plants to stand up, rather than lying on the ground as the mosses and liverworts did.

These first plants contained no roots. Roots, which dug into the ground to tap the groundwater's nutrients, were refinements to be developed later by the clubmosses, horsetails, and ferns, enabling them to build strong, woody stems that could grow to considerable heights. The plants also carried with them from their marine origin the capacity to use photosynthesis to create new body materials. To that end, green leaves were created filled with chloroplasts.

The combination of photosynthesis and roots carrying nutrient water enabled the horsetails and clubmosses to grow to immense heights of more than 100 feet, with trunks up to 7 feet in diameter. Most of our coal deposits come from giant forests of clubmosses, horsetails, and ferns laid down some 350 million years ago.

The evolution of land plants over the next 100 million years centered largely on developing sexual means of reproduction that did not depend on water as the medium in which the sperm cell would fertilize the egg cell. Water was a fine medium in the marine environment, but on land its relative absence posed distinct problems.

The first conifers—the ancestors of today's needle trees like spruces and pines—appeared about 100 million years after the first land plants emerged, that is, some 300 million years ago. Their present-day descendants—the pine tree, for example—rely on the wind to distribute their pollen (the sperm cells). When pollen lands on an egg-bearing cone, the pollen sprouts a tube that burrows its way down toward the egg cell in the cone. That process takes a whole year. When the tube has made contact with the egg cell, the sperm cell descends into the tube and fuses with the egg cell. During the second year after the pollen first landed on the egg-bearing cone, food deposits are laid down in the seed package in which the fertilized egg lies protected. An external waterproof coating also is laid over the seed, and then the cone dries up, falls down, and releases the seeds. Those cones can lie dormant on the ground for years, until moisture penetrates the protective coating and initiates the sprouting of the seed.

That astonishingly complex and ingenious solution to reproduction in a waterless, air-dominated environment took about 100 million years to be developed. It entailed an enormously complex degree of development on the structural level (pollen, pollen tubes, cones, seeds), on the level of cell differentiation, and on the molecular levels (in terms of new proteins, DNA codes, carbohydrates, and so on).

The development of the conifers and their complex cycle of sexual reproduction vastly compounds the complexity of life's development. At the same time, the period during which that development took place is severely compressed. (Recall that the development of a simple eukaryotic cell took over 2 billion years.) The Darwinian biologist, at this stage of accelerated evolution, now hides behind sex. Sex is chance's "magic wand" that purportedly accelerates such developments enormously, so that those astonishingly ingenious creations can take place. However, no biologist has yet to prove by hard mathematical calculations that, say, a million generations of protoconifers would produce the solutions just described for the conifers' reproductive cycle.

Doubts concerning the adequacy of the chance explanation crop up ever more as the inventiveness of Nature accelerates its pace. In shorter and shorter spans, more and more remarkable creations take place that culminate in the appearance of humans during the last 10 million years. This accelerating development is vividly depicted in Table 4.1.

From Table 4.1 we see that the magnificent invention of the flowering plant occurred about 120 million years ago. The origin of the flowering plants, or the angiosperms, was for Darwin "an abominable mystery" (E. C. Olson 1985, p. 10). Darwin's agonizing dilemma is not surprising, because the invention of flowers almost tangibly carries the imprint of an intelligence deciding to create them!

Two hundred million years earlier, the conifers developed a very viable mode of pollination by using the wind to help fertilize their egg-bearing cones. That mode has served very well for the conifers up until present times, and today conifers make up roughly one-third of all the forests on Earth. But it was not the most efficient mode of fertilization; the wind would haphazardly distribute the pollen, and only a small fraction would land on the egg-bearing cones.

A much more efficient way was to have the pollen delivered directly to the cone, say, by an insect. Flying insects seem to have appeared about 300 million years ago, having evolved from chitin-encapsulated crawling insects such as the ancestors of the bristletails, springtails, and silverfish. The first

BIOLOGICAL INVENTIONS

Table 4.1

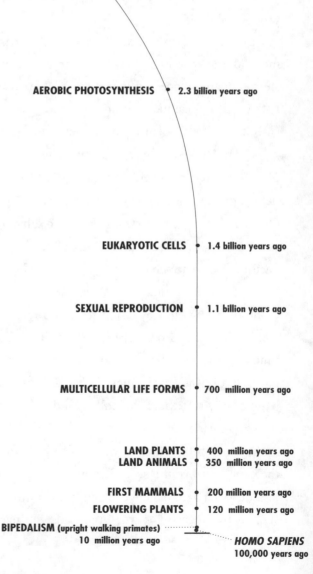

ORIGIN OF LIFE • 3.8 billion years ago

ANAEROBIC PHOTOSYNTHESIS • 3.1 billion years ago

AEROBIC PHOTOSYNTHESIS • 2.3 billion years ago

EUKARYOTIC CELLS • 1.4 billion years ago

SEXUAL REPRODUCTION • 1.1 billion years ago

MULTICELLULAR LIFE FORMS • 700 million years ago

LAND PLANTS • 400 million years ago
LAND ANIMALS • 350 million years ago

FIRST MAMMALS • 200 million years ago
FLOWERING PLANTS • 120 million years ago

BIPEDALISM (upright walking primates)
10 million years ago

HOMO SAPIENS
100,000 years ago

Life started some 3.8 billion years ago as primitive single-celled organisms. These reigned supreme in the oceans until the larger, more advanced single cells—the eukaryotic cells—appeared 2.4 billion years later. During this first period, anaerobic and aerobic photosynthesis were invented, which created an oxygen-rich atmosphere. Sex was invented soon after the eukaryotic cell appeared. Multicellular organisms began the invasion of the landmasses 400 million years ago. *Homo sapiens* is a biological upstart—invented only 100,000 years ago.

Source: Adapted from E.C. Olson, 1985, p.10.

flying insects were the dragonflies, which grew to the enormous size of 2 feet in wingspan.

As usual, the Darwinists can advance only lame suggestions as to the origin of winged insects. In David Attenborough's *Life on Earth* (p. 72), we encounter the suggestion that wings for flight first appeared as thin flaps on the body to increase blood circulation, thus increasing the insects' body temperature by exposing the flaps to sunshine. This heightened body temperature accelerated insect metabolism and made the insects more active. Wings, according to Attenborough, then formed "accidentally" out of these flaps and led to flight. To a non-Darwinist, however, it appears equally reasonable to hypothesize that *the idea of flight* could have occured to a kind of planetary intelligence and that this idea would then have been translated into rudimentary wing structures.

The spores and pollen were attractive food sources for many crawling insects, but up until the development of the flower, the pollen would usually develop in one place on a plant (or in a plant of different sex) and the egg-bearing organs would be in another place. For insects to pollinate the eggs, the male and female sex cells needed to be in the same place—which was accomplished in the construction of the flower.

If chance were the creative operative agent, we would expect that flowers would first appear in the conifer-related species, but that was not the case. The first flowering plant seems to have been the magnolia tree, which appeared out of nowhere—with its magnificently developed flower—a little more than 100 million years ago.

The sudden appearance of the flowering plants is an example of what paleobiologists call an "explosive radiation," with few preparatory signs and an extremely rapid diversification. Together with the insects, the flowering plants are by far the most abundant life forms on Earth today. There are about 200,000 species of flowering plants and about 700,000 insect species and there is a clear correlation between the rise of the insects and that of the flowering plants. It is as if the Creator found endless delight in spinning out thousands, yes millions, of varietal forms of those phylae. And what a variety has been created. Glorious colors to attract the insects, delicious nectars to be given as rewards, special pollen besides the fertilizing pollen, and delicious scents were all developed. Particular landing marks on some flowers were developed, especially visible in ultraviolet light, to show the insects where to land.

Let us consider two remarkable creations in the family of flowering plants that were developed to lure insects and to ensure that pollen sticks to them.

The first is the bucket orchid in Central America, which drugs its visitor. A bee that lands on that particular flower and sips its nectar will become tipsy and stumble about on a specially developed slippery surface inside the flower. The drunken bee slides down into a small container of liquid, and the only way out of that container is up through a specially formed spout. Inside the spout is an overhanging rod full of pollen that falls on the bee as it wriggles its way up. The pollen is then deposited at the egg cell of the next flower of the same kind that the bee goes rolling to.

The second creation is also from the orchid family in Central America and Ecuador. That orchid produces a flower that totally imitates the form of a female wasp, with eyes, antennas, and wings. It even gives off the odor of a female wasp under mating conditions. The imitation completely fools the male wasp, so he arrives for mating and receives a load of pollen that he then deposits in the next orchid flower that imitates a female wasp.

No wonder Darwin was confronted with an "abominable mystery." In the case of the orchid "wasp" design, the mystery is compounded not by the appearance of an arbitrary variation of any flower design but by a flower directly imitating the visual appearance of an existing *insect* design.

This is an extreme example of the many strategies adopted by various species either to defend themselves against other species or to attract the aid of other species to fulfill some needed function (in this case reproduction). We can truly speak of an interdependency in the evolutionary development of complementary species. In a later chapter we will touch on the larger interdependent web of ecosystems in connection with comments on Lovelock and Margulis's Gaia hypothesis.

So far we have dealt with two major evolutionary thrusts in multicellular life forms: those of the land plants and those of the insects. Two additional major creative advances were made at roughly the same time: the evolution of the fishes and subsequently the land animals.

In order to trace the ancestry of fishes—*and thereby ourselves*—we have to go back to the ancient era of worms. Already we have learned how something similar to the flatworm evolved into the crustaceans and the trilobites and that the insects evolved from the segmented worm.

Another of those ancestral worm forms began a different evolutionary breakthrough with the beginnings of an inner hard skeletal bone structure. We can see that ancestral form today in the very inconspicuous fish species called the lancelet, which appears as a slim leaf about 2 inches long. It usually lies half-buried in sand and receives its nutrients through an opening in the front. It has a few pulsating arteries, but still no fins and no limbs. However, through the length of the body there is a flexible but hard rod structure, a backbone of sorts to which are attached bands of muscles. The lancelet can rhythmically contract those muscles and thus propel itself through the water: it swims.

In a fascinating manner, we have a first crude draft of the design of a fish. Although the lancelet is active and alive today, there is no doubt that a very similar form existed some 550 million years ago. In the Burgess Shale in western Canada, among the fossilized worms, the forerunners of molluscs and crustaceans, lie fossil remnants of lancelet-type creatures. It is thus quite likely that the lancelet type of life represents the beginning of the vertebrates, the life forms having a hard, bony skeletal structure, out of which we humans developed some half a billion years later.

The lancelet type of life went on to evolve into the jawless fish represented today by the lampreys. Fossil remains of lampreylike fish have been found dating back 540 million years, a little younger than the age of the Burgess Shale.

Lampreys have two well-developed eyes. The slits in their sides, which had acted in the earlier forms as filtering mechanisms, were walled in with blood vessels, so they also served as gills with which to take in oxygen from the water. The lampreys also have a well-developed tail fin, which enables them to swim. Their mouth is jawless and consists of a disk-shaped structure containing a tongue with sharp spines, with which the lampreys eat dead or live fish. The fossilized version of the lamprey had only a tail fin, so it could not swim high above the sea floor. And without a jaw, it could not attack food sources such as molluscs. However, the lampreylike ancestral fish represents a second draft of a design for a real fish.

The jawless fish were small, like minnows, and heavily plated with an external bone structure in addition to their inner skeletal backbone. Inside the bone structure of their heads, the beginning of a brain much like that of modern lampreys was formed. An intriguing balancing structure had also been developed in the form of two arching tubes that projected out horizontally to each side. The motion of liquid inside the tubes against the pressure-sensitive inner surface enabled the jawless fish to balance in the water. That device thus served as a kind of carpenter's leveler—an amazingly sophisticated device at such an early stage of fish development. The same tool is found today in the living lampreys. That tool later developed into sophisticated ears for hearing underwater.

The next important find among the fossilized fish dates back to about 400 million years ago. It was located in the northwestern part of Australia at a place called Gogo by the aborigines. There, in the sedimentary rocks, the fossil remains of the first true jawed fish have been found. A large variety of jawed-fish fossils were found there. They all show heavily toothed jaws that developed from the anterior pairs of gill slits, as well as an emerging vertebral column that surrounds the flexible rod structure of the back. They also show well-developed lateral fins, mostly in pairs, near the head and two near the anus, in addition to the tail fin. Some of those jawed fish were bottom dwellers, others were free-swimming. Two types were gigantic, more than 18 feet long.

Such was the evolutionary situation for the fish 400 million years ago. At that time, some of the marine plants were beginning to move onto land, and the first mosses and liverworts began to appear. That was also the time when the first insects, such as the millipedes, began to crawl onto land. It would take another 50 million years before the first amphibian fish would begin to move onto dry ground. However, during those 50 million years— an incredibly short time span, evolutionarily speaking—there was feverish activity in fish development.

During that time, a momentous split occurred in the further development of the jawed fish. Two major groups, the rayfinned fish and the lobefinned fish, formed. The rayfinned fish were the ancestors of all living fish of today, whereas the lobefinned fish were the ancestors of the amphibians, which ventured onto land, where the plants, having arrived earlier, served as food sources.

During the further development of the rayfinned fish, the creation of the swim bladder represented a very clever evolutionary breakthrough. By diffusing oxygen gas into or out of the bladder, a fish could regulate the height or the depth in the water in which it wanted to swim. Fish had acquired variable buoyancy. They could use their fins not to keep afloat but solely for the purpose of propulsion or locomotion. And thus the marvelously streamlined, hydrodynamically efficient bodies of the tuna, the bonito, and the mackerel (the speed swimmers of the lot) were developed.

The evolutionary breakthrough in the lobefish was the development of bony structures inside their front pair of fins. The fossil records clearly show that development to be the beginning of the amphibians. Remarkably enough, there persists a species of fish today that shows exactly the development of bony flipper fins as evinced in the fossil records: the coelacanth fish. Fossils of coelacanths are abundant from 400 million years ago, and they persist in the fossil records until about 70 million years ago. Today such fish are found regularly living in the oceans by the Comoro Islands in the Indian Ocean midway between Madagascar and Tanzania, in the same shape as the last fossils deposited 70 million years ago. It caused a sensation in the scientific world when the first living coelacanth was caught in 1938 off the coast of South Africa. That fish showed exactly the same type of bony flippers the fossil records revealed.

Survival on land required the development of new tools. Bony limb structures were needed to support the weight of the animal on ground where water buoyancy is gone. Another requirement was equipment to breathe air—some form of lungs.

It turned out that the swimming bladder, which served in the rayfinned fish as a buoyancy tool, could be adapted into primitive lungs. All that was needed was to line the bladder with blood vessels. The blood vessels could then absorb the oxygen from the air gulped down by the fish breaking the surface of the water. That mechanism is still possessed by the bichir fish today, which lives in the rivers and swamps of Africa. The bichir uses both gills to get oxygen from the water, but it also periodically gulps air, which goes into a pouch converted from an ancient swimming bladder. Those same swamps and rivers in Africa are also inhabited by the lungfish, which has developed a pair of pouches to serve as lungs together with the gills.

The fossils of ancient lungfish have been found in sedimentary rocks in Africa, Australia, and South America. Those rocks all date back to 350 million years ago, and they also show fossils of the coelacanths. If we recall that the first jawed fish appeared 400 million years ago, that leaves only 50 to 70 million years between the first decent fish design and the appearance of the amphibians. That is, evolutionarily speaking, an incredibly brief time for chance to enable such momentous developments. And we must of course bear in mind that such evolutionary steps are traced down to the level of DNA and the assorted new molecules.

According to the Darwinists, it is not enough to create a successful protein, then a new cell type, and then cell structures to produce the first bony fins and the first rudimentary lung; these steps have to be done *accidentally*—nudged by natural selection. If chance were somehow *aware* of the idea of a crawling limb or a lung, such developments would be quicker to evolve, but chance is not aware. Chance has to play blindfolded to stumble upon the notion of manufacturing bones in a pair of frontal fins or converting a swimming bladder into a pair of lungs. Darwinians would have to say that 50 million years is sufficient time for such elaborate designs to come into being by chance. But it took chance far longer—700 million years—simply to create the first multicellular animals by using the tools of sex and eukaryotic cells.

In the African fossils formed 350 million years ago are found remnants of an amphibian that combines the bony frontal limbs of the coelacanth with the rudimentary lungs of the lungfish. The head of that fossilized creature shows the first connecting passage for breathing between the nose and the roof of the mouth, a fundamental skeletal feature of all land vertebrates. Thus, that fossilized life form, named Eusthenopteron, can legitimately be considered the ancestor of the first true amphibians, which crawled onto land but for millions of years alternated between a watery environment and a dry land environment. Such a transition may seem gradual, but in evolutionary terms it was relatively fast.

Biologist David Attenborough muses in *Life on Earth,* "Why should the descendants of Eusthenopteron have troubled to clamber about laboriously on land?" And he—representing the traditional biologist—is not able to give a clean, straight, unequivocal answer. *Maybe* the animals were in search of new food sources, or *maybe* they lived in isolated seasonal pools and used their legs and lungs to seek new places of water when the old ones ran dry. Time and again, when pressed for a hard scientific answer, the Darwinist resorts to handwaving, giving evasive qualitative answers never supported by quantitative calculations in mathematical probabilistic terms.

On the other hand, everything falls into a meaningful pattern if we look at evolutionary developments within the framework of a creative intelligence that is growing, experimenting, learning, and developing. Viewed in this context, the movement onto land by the amphibians makes sense; land becomes a new laboratory in which the intelligence can experiment and "play."

And play it did. In a very short time there evolved amphibian life forms that roamed dry land. Some grew into formidably large structures 10 to 12 feet long, with huge jaws spiked with rows of cone-shaped teeth. For almost 100 million years they dominated on land. Modern descendants of those amphibian forms live on in purest forms in the salamanders, newts, and frogs.

However, if we look at the amphibians as an experiment, they were a good beginning, but they could not serve as a fully viable land animal.

Although amphibians had solved the problems of locomotion and air breathing, they were still far too dependent on water for reproduction and habitat.

The next two very important steps in the evolution of the amphibians were the inventions of the watertight skin and the encapsulated shelled egg. The development of the watertight skin also led to the invention of a "thermostated" body. The first life forms to be endowed with those two inventions were the reptiles, which appeared almost 280 million years ago. Again, we note the rapidity with which such important and complex developments were made. The reptiles appeared only 70 million years after the first true amphibians.

The watertight skin and the encapsulated egg proved very successful, and the reptiles rapidly developed into two successful groups, the pelycosauruses and the therapsids. Again we see the tentative, probing steps as evolution unfolds. For 60 million years the pelycosaurus reptiles dominated on land. Then the therapsids took over, and from them the first mammals evolved. It is as if the attention of the creative intelligence vacillated or followed impulses—for 60 million years the pelycosaurus experiment took priority, and then attention turned to the therapsids.

Independently of the pelycosauruses and the therapsids, however, the dinosaurs appeared. They represented a branch that had laid dormant during the reign of the pelycosauruses and the therapsids. But then the mammal branch of the therapsids began to lie dormant while the dinosaur branch dominated on land for the relatively long period of 150 million years.

The dinosaurs seemingly did not add any evolutionary breakthroughs; they merely brought to full flowering the use of the watertight skin and the encapsulated egg. Their thermostat device, however, was only a first draft. The dinosaurs were ectothermic (cold-blooded); that is, they gained heat from their external surroundings rather than generating body heat internally like the endothermic (warm-blooded) mammals. (Some biologists argue, however, that the dinosaurs may indeed have been warm-blooded creatures.)

All living reptiles, such as snakes, lizards, and the Galapagos iguanas, are called cold-blooded, but in reality the body temperature of an iguana is the same as that of a human being and in some lizards may even be a few degrees higher. The great advantage of maintaining a relatively high body temperature is that the physiological processes, which involve chemical reactions, proceed faster, thus generating more energy.

The ectotherms' externally conditioned heat supply and thermostating require much less energy than the endotherms. About 80 percent of the calories of the food intake of the endotherms (mammals) are used to generate body heat. An ectotherm can survive on only 10 percent of the mammalian nourishment, a fact that allows reptiles to survive in deserts. On the other hand, the great advantage of generating your own body heat is survival in cold climates and physical activity during the night. This advantage proved crucial in the further evolution of life on Earth.

The fossil of the first undisputed mammal, found only about twenty years ago in South Africa, dates back as far as 200 million years ago. This animal is only 4 inches long. Its jaw and the skull, together with its teeth, link it to something like the shrew of today. In a sense, it was a premature development, because the dinosaurs totally dominated the land continents at that time.

It took about 135 million years before the mammals could really begin to flourish, when the climate got markedly colder 65 million years ago and the dinosaurs finally became extinct. During the dinosaurs' reign, the mammals perfected the endothermic thermostat, for which a large food intake was necessary. Accordingly, refinements in teeth structures were made to allow mammals to grind and chew or tear food. The endothermic system permitted day-and-night operation of delicate and complex organs requiring stable temperature conditions inside the body.

With the extinction of the dinosaurs, however, the land lay free for the mammals to explore. Some of the mammals took to the trees, a momentous move because it led to the development of some very useful tools: a pair of forward-facing eyes and hands with grasping fingers. Those tools do not clearly represent evolutionary breakthroughs; rather, they were adaptations of previous tools to the new environment.

Likewise, when a particular group of primates, the apes, began to descend from the trees and walk about on their two hind legs, that bipedalism appeared as a useful adaptation. We could say that the drafting of humans began some 50 to 65 million years ago but that the evolutionary breakthrough happened only 8 to 10 million years ago, when the brain of a group of apes (or apelike ancestors) began to grow far out of proportion to that animal's need for survival.

Evidence indicates that the first mammals coming out of the reptilian line continued creating encapsulated eggs for reproduction, but with certain important modifications. We observe relics of this epoch today in two strange animals: the water-loving platypus and the ant-eating echnida in Australia (Attenborough 1979, p. 204). They both lay eggs, although the echnida places the eggs in a pouch to hatch, and they both produce milk to feed their babies. Unfortunately, there are no fossil records to trace the direct ancestry of the platypus and the echnida, but their reptilian mechanisms suggest that they are living fossils in the sense of presenting a transitional stage between reptiles and mammals in the evolution of reproductive cycles.

The mammals later began to develop the organs for replacing the shelled-egg invention of the reptiles. We see that shift in the development of the marsupials, whose offspring gestate inside the mother until birth. After being nourished by an internal yolk sac, the young ones have to wriggle out through the cloaca passage of the mother and settle in a pouch on the stomach, where they drink milk from teats.

Secondary characteristics that developed include a furry coating to cut down heat losses, a respiratory diaphragm, a muscular structure that greatly improved agility, and facial muscles that allowed suckling at the mother's teats. We may indeed speak of those developments as some of the last evolutionary prerequisites to the development of the placental method of reproduction.

The need for developing more and more complex organs for breeding the next generation resulted from the increasing number of complex organs that needed to be created within the embryos. Egg yolks could not provide a sufficient amount of nutrients. To give the fetus time to develop its complete body necessitated a protected environment with plenty of food.

Thus, the astonishingly complex structure of the placenta was invented. The placenta is a flat disk that is connected to the uterine wall on one side and to the umbilical cord of the fetus on the other side. The placental tissues connected to the uterine wall are very convoluted, so that the two surfaces share a relatively large area. This facilitates the diffusion of oxygen and nutrients from the blood of the mother in the uterine wall to the blood of the fetus in the placenta. Likewise, the waste products of the fetus diffuse into the mother's blood to be excreted through her kidneys. In other words, no blood is exchanged between the mother and the fetus.

At this time we should reconsider the preceding paragraph and ponder the complexity of this information from the point of view of such details being created simply by chance. What really matters is to explain how such numerous functions within the organ could be orchestrated on the lower levels of organization into a harmonious, integrated whole—the placenta.

And there exist added refinements. The placenta secretes a hormone that suspends the mother's sexual cycle for as long as the placenta is in place, so that no more eggs are produced to compete with the fetus in the uterus. In addition, the fetal tissues are not genetically the same as the mother's; they contain elements from the father. Thus, when it becomes connected to the mother's body, the fetus risks immunological rejection, in the same way a transplant does. Just how the placenta prevents that from happening we still do not know in detail, but it seems to be regulated by the production of other hormonal substances.

The enormously complicated interplay of subordinate organic structures and chemical secretions that produces the placental method of reproduction must have been developed during probably 50 million to 100 million years. For chance, by accidental mutation (tested by the natural selection of the environment), to develop such a superbly complex orchestration of tissues, hormones, blood nutrients, waste transportation, and a mechanism to sidestep immunological rejection is just not mathematically possible. It is simply not imaginable!

Every new turn of evolutionary breakthroughs is more and more complex and occurs in a shorter and shorter time span. When we take this fact into account, the outlandishness of the chance hypothesis becomes more and more flagrant and improbable.

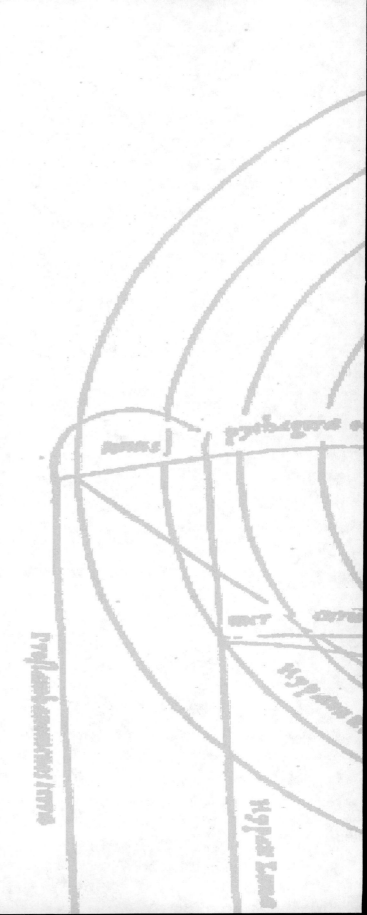

The "ALL-SEEING" EYE

P eople, the poetic thinking animals that they are, have termed the eye "the mirror of the soul." That lovely expression focuses on the tendency of the human eye to reveal a person's inner being. For most other life forms, however, the eye is simply a means to look out into their environments. The eye or eyespot is used to search for food and to detect enemies. Just as limbs were among the earliest appendages in multicellular or even single-celled life forms, so too were there some single cells that specialized in reacting to light. The development of the light-sensitive cell had as its consequence the development of a data-processing center, a nerve center from which signals would go out to muscles to react to light stimulation.

The origin and development of the eye is an evolutionary breakthrough of the first order. It is hard to decide which of the five senses in human beings is the most important, but without the eye our perception of the Universe would be close to naught.

The details of the evolution of the eye are difficult to map because the eye is a soft body that does not fossilize well. Still, we must marvel at the inventiveness and the apparent intelligence by which various species have become equipped with optical systems serving their special needs. In the words of Lyall Watson, in his book *Lifetide,* "Darwin himself admitted that the perfection of the vertebrate eye, sent cold shivers down his spine." As Watson puts it:

> The transparent cornea of our eye could hardly have evolved through progressive trial and errors by natural selection. You can either see through it or you can't. Such an innovation has to be right the first time or else it just does not happen again, because the blind owner gets eaten. (P. 166)

However, regardless of their causative agents, perfected seeing devices have developed in four families of life forms: the fish, the squids, the insects, and the vertebrates, including the mammals. In each family a wide variety of sophisticated eyes has been developed.

Exactly when the eye first developed is difficult to ascertain. It is, however, significant to note that in the Burgess Shale, laid down about 550 million years ago, the lancelet appears to be the ancestor of the fish, and the trilobite appears to be the ancestor of the insects. The lancelet as we know it today has no eye, only a light-sensitive spot. The fossilized trilobites show five eyes, one directed upward and the other four having side views. If multicellular life forms first appeared some 700 million years ago, the first eye must have been invented over a period of a mere 100 million to 150 million years, an astoundingly short period of time for such a complex development.

The light-sensitive spot persists in some life forms today in the form of an ocellus (meaning "little eye"), which has no lens to form an image. Some flatworms, jellyfish, starfish, and insects have a pair of ocelli, bulging cup-shaped spots made up of a cluster of pigment cells with an opening. Inside the cup lie the endings of special nerve cells called retinula cells. The function of the dark pigment cells is to shield the retinal cells from light coming from somewhere other than the opening of the cup. The tips of the retinula cells, called the retinula clubs, contain photosensitive molecules in a visual pigment. Light striking those retinula clubs generates a nerve impulse, which is then registered by the rudimentary brain or nerve center.

That configuration is the first crude draft for the optochemical design of the eye. During several tens of millions of years, refinements such as lenses and mirrors were added. But the eye retained the principle of detecting light by means of retinal cells and then sending nerve impulses to a central processor, the brain.

The fish, which evolved from a different branch of the tree of life than the trilobites, apparently developed complete eye structures independently, at almost the same time as the appearance of the trilobites some 550 million years ago.

Whereas the multiple eyes of the trilobites eventually evolved into the mosaic eye of the insects, the fish evolved paired eyes for vision that protected their sides. Early on there also appeared a third eye on top of their heads to protect them from attacks from above. That third eye remains as a rudimentary form in some now-living fish, but it is incompletely developed. It indicates light or darkness but forms no image because it lies buried under the skin. Still, it comes complete with a lens and retina. In humans, the third eye survives as a small stalked red body, lying buried between the hemispheres of the brain. The famous French philosopher and mathematician René Descartes (1596–1650) thought that this glandlike body, *glandula pinealis,* was the seat of the soul.

Over the course of 500 million years, fish have developed some seeing devices that have remarkable and amusing functions. There exists in Central America a fish, the anableps, that sports four eyes. The four are in fact two eyes having a horizontal division, so that the upper half of the eye pair sees in air and the lower half sees in water. Thus, the fish can hunt for food both above and below water. Then there is the archer fish, which has a pair of eyes that compensate for viewing objects in the air from underwater. Those fish prey on insects in the air, and upon finding one suitable for eating, they shoot a jet of water up into the air to knock the insect into the water.

Fish living near the surface of the water have color vision, whereas others, like the shark, do not see colors. Then there are the mechanical constructions in the eyes of octopuses, squid, and amphibians, such that the eye lens focuses like a camera by moving forward or backward to accommodate different distances.

In some ways the eyes of the squid are better than those of the human. Their huge size (up to 16 inches in diameter) permits the squid to resolve finer details than the human eye, and the squid eye, like the eye of the bee, responds to a quality of light called polarization, which probably helps the squid navigate.

The camera-lens eye construction of the mammals and their ancestors, the fish, derives from the light-sensitive spots of the flatworms. So do those of the squids and trilobites. The light-sensitive spots of those flatworms that evolved into fish, molluscs, and squids led to almost identical approaches to vision: a variable-shaped lens, an iris, a retina, vitreous substance, and optical nerve cells connecting to a brain. According to the Darwinists, in two totally different evolutionary paths (the fish and the molluscs), chance interacting with natural selection invented the same seeing device. That notion is almost absurd, unless the *idea* of an eye occurs first and is acted on to solve the problem of vision in two altogether different lines of evolutionary development.

Let us consider next the eyes of the insect world. As stated earlier, the insect eye evolved out of the trilobite eye. The trilobite eye represented an amazingly advanced optical design for such an early evolutionary stage of multicellular life forms. The trilobites were the first to develop "high-definition" eyes by orienting up to 15,000 crystalline lenses, which also gave them hemispheric (or 180-degree) vision, something humans do not have. Furthermore, as a later development in their reign of about 300 million years, some species of trilobites evolved a unique development that human civilization developed only 100 years ago: the doublet lens, which is used for vision underwater at low light levels. In such circumstances a thick convex (outward-bulging) lens is needed, but this will not give a sharp image unless a concave (inward-bulging) lens is added. That combination is exactly what those trilobites developed. Their eyes had lower and upper lens elements, and the surface between the two conformed to a mathematical optical principle that humans discovered as they tried to correct spherical and color aberrations of lenses in telescopes. (We refer to this configuration as the Fraunhofer lens system.)

Thus, the trilobite's eye required optical engineering knowledge of considerable sophistication. Can chance accidentally hit upon knowledge such as this? It is hard enough to envision a single lens being built by chance.

Nature has constructed many other incredibly imaginative solutions to eyes. Among the many contributions of Sir Isaac Newton to science was the invention of the pinhole camera and the mirror telescope known as the reflector. But even those devices were long before discovered and used by Nature. The pinhole camera is a box with a small aperture (a pinhole) in it. Thus, it uses no lenses. The eye of the Nautilus mollusc with flotation chambers operates using this principle. Instead of placing film where the image is formed, Nature placed light-sensitive retinal cells over a curved area. Such cells use the photoelectric effect, which human astronomers began to use only fifty years ago. With its pinhole-camera eye, the Nautilus mollusc receives a picture upside down on its curved retinal area. Its rudimentary brain, which receives that image, must have an image-processing central network to allow the animal to interpret the image as right side up.

In the other molluscs, such as the scallop, Nature has experimented with combining mirrors and lenses, yielding in some ways an even more extraordinary optical system than the double-lens system of the trilobites. The scallop displays about fifty eyes when its shell is opened. Each of those eyes is a compound optical system made up of a parabolic mirror and a correction lens.

That design represents an extraordinarily sophisticated optical system. It resembles a telescope design that was invented by an optician named Bernhard Schmidt just before World War II. In the Schmidt telescope, a spherical mirror is used along with a lenslike correcting glass plate placed in front of the mirror. If the correction plate has the proper intricately shaped curve, its combination with the mirror will yield a wide field of view sharply in focus. The Schmidt telescope was considered a revolutionary advance in the human field of optical design. It is still of great value in survey work of astronomical objects. However, Nature has been able to use a *parabolic* mirror and a correcting lens to provide a wide field of sharp vision in the scallop eye. It is considerably harder to produce a parabolic surface than a spherical surface.

The mirror-lens solution, however, has not been adopted as the major design for seeing devices for most multicellular life forms. It is as though Nature experimented with the concept and tossed off a few special designs before settling down to serious business in the designs of the insect eye and the vertebrate eye. It is hard to say which of those two "serious" designs is the more glorious.

The basic element of the insect's compound eye is the ommatidium, the "little eye." The number of little eyes varies greatly from one insect species to another. The ant, for example, has only 6 ommatidia, whereas the common housefly has about 4,000 and the dragonfly a whopping 28,000. Most flying insects have at least partially binocular vision (using two "complete" eyes), as well as a 360-degree field of view. (However, groundhuggers like the beetle need only monocular lateral vision.) The all-around vision of a fly is the reason you cannot sneak up on it from behind to swat it.

Each ommatidium is a self-contained optical system. It has two lenses in a tube that leads to a cylinder in which are placed six to eight neuron cells, called retinula cells. From those cells, neural axons lead directly into the insect brain. Insects have no optical nerve bundle as in the vertebrate eye.

The ommatidia are hexagonally shaped, like the cells in a bee's honeycomb. The hexagonal shape provides the most economical use of an existing surface area, since there is no wasted area in the boundaries between adjacent hexagons. In the construction of those little eyes, the insect's body armor (the cuticle chitin) is changed in the compound eye area to become transparent. The transparent cuticle forms the outer corneal lens (see Figure 5.1). Inside that outer lens lies the crystalline cone (another lens), which is not a hard, dead protein like the corneal lens but is made up of living cells. Those two lenses sit encased in a cone where the walls are coated with

INSECT EYE CELLS

Figure 5.1

The retinula cell membrane is highly elongated and cylindrical in shape. Along one side it is folded back and forth to form the microvilli. These folds are coated with a light-sensitive molecule called rhodopsin. Upon receiving a single photon of light, the rhodopsin molecules at the upper end of the retinula cell initiate a chain reaction of molecular activities that results in an amplified electric signal exiting from the lower end. This electrical signal propagates via nerve cells into the vision center of the insect brain.

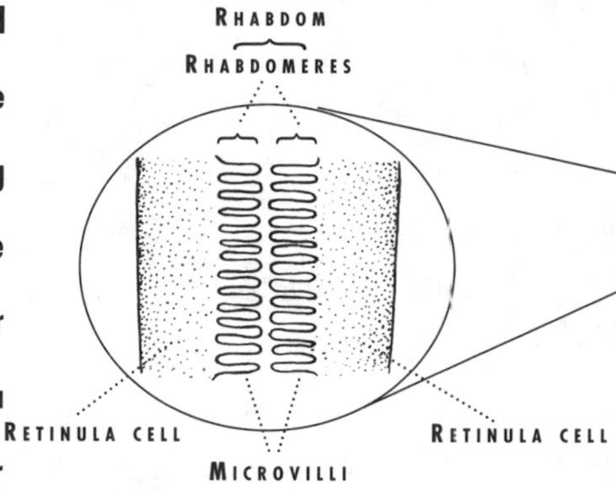

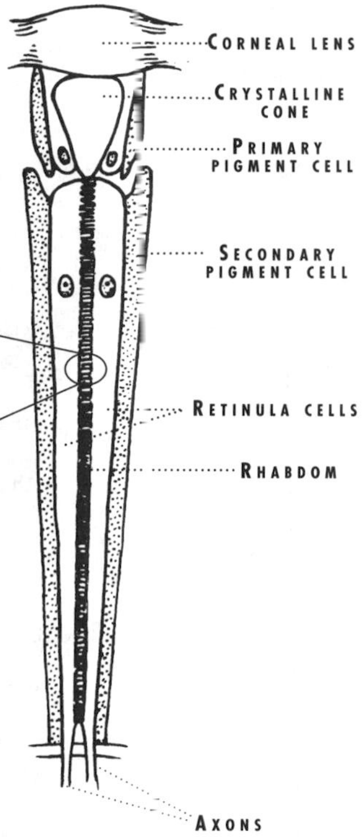

The same principle of light detection is utilized in most insects and animals. In the human eye, there are over 100 million retinula cells (rods and cones) that form the retina. The human eye is the most sensitive light detector ever devised—it is able to react to a signal as small as one photon and to one as great as a trillion photons. By contrast, an electronic photomultiplier, the imitation of the retinula cell constructed by humans, responds to a range of one photon up to only a few million photons.

RHABDOM
RHABDOMERES

RETINULA CELL
MICROVILLI
RETINULA CELL

CORNEAL LENS
CRYSTALLINE CONE
PRIMARY PIGMENT CELL
SECONDARY PIGMENT CELL
RETINULA CELLS
RHABDOM
AXONS

Above is depicted a cross section of the optical design of the individual eye element of an insect—the ommatidium. In the dragonfly, 28,000 such elements form two composite eyes, each with 180-degree vision. Light from the individual eye lens falls onto the front ends (rhabdomeres) of retinula cells (see above left).

dark pigment cells. Likewise the retinula cells are surrounded by dark pigment cells. The function of the pigment cells is to prevent light from being reflected from the sides of the cone onto the photoreceptive part of the retinula cells. (In astronomy, a similar method, called "baffling" the telescope, is used to minimize the extraneous, scattered light onto the image of the object being observed, in order to preserve the inherent contrast in that object.)

The six to eight retinula cells combine their photoreceptive parts along a central axis in a structure called the rhabdom. Along that axis the cell membrane of each retinula cell folds back and forth to form the microvilli, which ensure that the light coming through the lens system along the axis will activate at least one light-sensitive molecule. These microvilli folds are all coated with the light-sensitive molecules; thus, they are the photoreceptors.

The optical system and the data-processing system just described represent what is called the apposition mode of the compound insect eye. Each little eye, with its small field of view, deals with a small part of the image of the object being viewed. Together, the ommatidia in the apposition type of compound eye form a mosaic image of the object. (Incidentally, insects read out their image data much faster than humans. People can differentiate 16 images per second, whereas the common housefly can differentiate 200 per second.)

The apposition system of the compound eye is used by insects that thrive in the daytime. We might think that such a mosaic image would be lacking in sharpness, but sharpness really depends on how small and how close together the image points are. A newspaper photograph is made up of thousands of tiny dots, which from a distance give the reader the impression of a sharp image. Thus, the fly (with 4,000 ommatidia) and especially the dragonfly (with 28,000) have perfectly sharp vision. The compound eye does not adjust for focus and does not need to, because it builds up a mosaic image of dots of varying brightness. (See Figure 5.2.)

For insects operating during nighttime at low levels of light, a different vision system was invented, called the superposition compound eye. The superposition optical system has hundreds or thousands of individual

THE FLY EYE

Figure 5.2

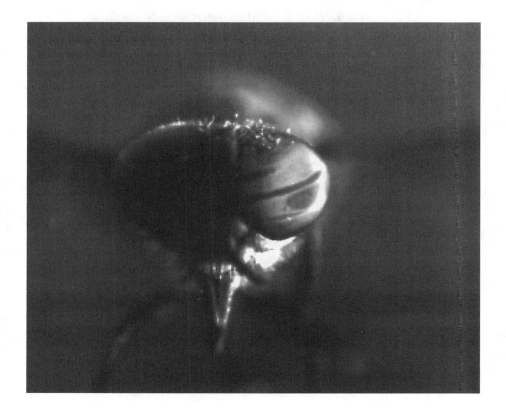

The common housefly has two compound hemispheric eye units, each made up of 2,000 individual eye lens units, or ommatidia. Light from an outside source is focused through the ommatidium onto the photosensitive retinula cell. Thousands of retinula cell/nerve cell connections have been precisely wired to give the perception of a well-defined image in the fly brain. Humans can read out 16 images per second; the housefly can read out 200 images per second.

ommatidia, as in the apposition eye, but the retinula cells do not touch the crystal cone. Instead, the cells are some distance away from the cone, connected by a tubular filament extension of the cone's membrane. The dark pigment cells are concentrated in the area near the crystal cone so that there is a region of transparency between the end of the cone and the retinula cells. That arrangement permits light from one point of the object to be reflected through the optical systems of *several* ommatidia and onto *one* rhabdom (neural photoreceptor element). Thus, a single point of the object is imaged onto one rhabdom through the superposition of individual images from the several ommatidia. Some nocturnal insects superpose up to thirty ommatidia images onto one rhabdom, which means a thirty-fold increase in the light sensitivity of their eyes.

It is absolutely fascinating to contrast the inventiveness of Nature with that of modern human beings. In the last ten years, instrumental astronomers have gotten very excited about three technological developments. One is the development of panoramic detectors called coupled charge detectors (CCDs), which are a mosaic of tiny detectors arranged in a square. Typically an array has 65,356 (256 × 256) detectors, only slightly more than the number of photodetectors in a pair of dragonfly eyes, whereas in the human eye the number of photodetectors reaches hundreds of millions. The second development that excites some astronomers is the use of fiber optics to transfer astronomical optical images certain distances to the photoreceptors. Nature already has such an invention in the form of the superposition compound eye, with the filamentary extension of the retinula cells. Furthermore, the very method of superpositing optical images to increase light sensitivity is now being touted as a "revolutionary" new telescope design in the so-called multimirror telescope in Arizona. The new Very Large Telescope configuration for European astronomers at the European Southern Observatory in Chile, their "flagship" telescope for the 1990s, uses the superposition principle invented by Nature several hundred million years ago. Four 8-meter-diameter mirror telescopes combine their output of light from one source.

Nature still has many more tricks in her bag of insect eye design. In some insect eyes, at the bottom of the retinula cells, part of the cell membrane is a reflecting mirror called the *tapetum lucidum*. A light particle that somehow is not trapped in a photoreceptor molecule on the way down through the rhabdom gets a second chance to be captured as it is reflected back up. Some other insect eyes have built-in mechanisms to switch from the super-position mode of operation at night to the apposition mode in the daytime.

However, the vertebrate eye, in particular the human eye, is the supreme seeing device created by Nature. In many ways, humans still have not been able to duplicate the entire system. In the vertebrate eye, for example, the lens can change its shape, so that it has a variable focus to see clearly whether the object in view is far or near. The eye lens in fish and squids moves back and forth like a camera lens to achieve the same result.

The human optical system involves a large number of highly differenti-ated cells. Three layers of different tissue form the eyeball. First there is an outer layer of tough tissue called the sclera, which is the white of the eye, except in front, where it is transparent and forms the cornea. The function of the sclera is to withstand pressure from within and without. In addition, the cornea lets light into the interior of the eyeball. The second layer, the choroid, is the front middle layer of wall tissue, which in the front forms the iris and the ciliary body that support the lens. Because the eye is made up of living cells (complete with mitochondria inside each cell), it needs nutrients to live. The choroid tissue brings those nutrients. The iris may be blue, green, or brown, depending on the color pigment provided for by the genes. The iris color pigment absorbs light, thus protecting the retina from exces-sive light. In the center of the iris lies the clear pupil entrance. The eye lens also consists of living cells and has a yellow color that absorbs ultraviolet light. The third and innermost layer of the eye wall constitutes the retina, the mat of photoreceptors called rods and cones. The receptors lead to the optical nerve, which transmits the signals that light generates to the brain Between the cornea and the iris is a chamber filled with fluid, the aqueous humor, and behind the lens is another cavity filled with a more jellylike sub-stance, the vitreous humour.

The eye is not a simple single-lens system. It is an optically complex two-component system. The cornea, the outer transparent shell, functions as a lens and brings an image to focus near the retina. The eye lens constitutes the second component, which together with the cornea forms a sharp image onto the retina. If we had only the cornea, our vision would be fuzzy. The fixed surface of the cornea and the variably curved surface of the eye lens together provide a sharp vision of distant or nearby objects as close as 10 inches without needing to change the distance between the lens and the retina.

That extraordinarily versatile but complex optical system works with three pairs of muscles, respectively, that move the pupil up and down, move it sideways right and left, and slightly rotate it. The ciliary body—the middle layer of tissues in the eyeball—contains the suspensory ligaments of the eye lens. When the ligaments relax the lens bulges, bringing near objects into focus. When the suspensory ligaments tighten and pull, the lens shape becomes more slender and brings distant objects into focus. In addition, dilator and sphincter muscles open and close the pupil size in response to different amounts of light. That design is completely analogous to the workings of a camera but is done with living cells and is automatically regulated by the brain. As we so often find in our beautifully constructed human form, simple actions that we take for granted, like focusing our eyesight on near or distant objects, involve a vast number of secondary processes to achieve.

The light-receiving and initial data-processing system in the eye together form the retina. But unlike the astronomer's panoramic CCD detector systems, Nature places the data-processing system of neural cells *in front* of the photoreceptors (the rod and cone cells). Light must pass through most of the retina before it reaches the photoreceptors, but that does not matter because that part of the retina is transparent. In back, near the choroid, lies the pigment epithelium cells that contain the dark pigment melanin. The function of melanin is to absorb light that does not hit the photoreceptors, so that light does not bounce around in the eyeball. Bouncing would produce scattered light, which disturbs the signal. When we look into a pupil, we see that it is totally dark. Astronomers also paint their telescope interiors with a special black paint, but Nature does the job better.

In the human retina, there are about 100 million rods and cones (Jastrow 1981, p. 82), whereas the best CCD devices commonly used by astronomers today contain about 1 million individual detectors. The human-designed single detector, the silicon diode, can respond to light over a range of about 300 to 1 in intensity, whereas the human eye spans a range of 1 trillion to 1. The human eye can see both in bright sunlight and on a starlit night when the illumination is a trillion times fainter (Alberts et al. 1989, p. 1101). (Common camera film can tolerate perhaps a factor of 10 in light intensity before it is overexposed.) The eye works at room temperature, whereas astronomers must cool their CCD devices to the temperature of liquid nitrogen (–196 degrees centigrade) to detect a few photons.

The multicellular structures in the eye are indeed astoundingly ingenious. But on the deeper molecular level of enzymes, proteins, and the like, there is a whole new dimension of complexity that the human mind has barely begun to comprehend. We will undoubtedly increase our understanding in due time, as exemplified by the spectacular successes already achieved in microbiology. But every deeper level of penetration discloses an ever larger degree of complexity, for which an inconceivable number of parts must function together as an integrated whole.

Darwin himself was very well aware of the difficulties his theory of natural selection would have in explaining the details of so exquisite a sensory tool as the eye. In fact, as a testimony to his remarkable intellectual honesty, he wrote a special chapter in his *On the Origin of Species* entitled "Difficulties of the Theory," wherein he writes with obvious admiration for the beauty of Nature's design:

> To suppose that the eye with all its inimitable contrivances for adjusting the focus to different distances, for admitting different amounts of light, and for the correction of spherical and chromatic aberrations, could have been formed by natural selection seems, I freely confess, absurd in the highest degree.

But Darwin goes on heroically arguing for the success of his principle of natural selection:

> It is scarcely possible to avoid comparing the eye with a telescope. We know that this instrument has been perfected by the long continued efforts of the highest human intellects, and we naturally infer that the eye has formed by an analogous process. . . . Further we must suppose that there is a power, represented by natural selection, or the survival of the fittest, always watching each slight alteration in them, preserving each which in any way or in any degree tends to produce a *distincter image*. Suppose each new state of the instrument to be multiplied by the million; each to be preserved until a better one is produced, the old ones all to be destroyed. . . . Variations will cause the slight alterations, will multiply them almost infinitely, natural selection will pick out with unerring skill each improvement. Let this process go on for millions of years; during each year on millions of individuals. . . . *May we not believe that a living optical instrument might thus be formed superior to one of glass?* [Italics added]

We must notice here the complete absence of mathematical formulations in terms of probability theory. Darwin merely alludes to the obvious occurrence of the process of natural selection operating on "millions of individuals" during millions of years. And he ends that loose reasoning with an appeal to the reader's credulity in a sentence that begins with the words "May we not believe. . . ."

By the parallels drawn between Nature's eye and the tools of the modern astronomer, we have seen that the simple function of seeing requires a vast body of sophisticated technical knowledge. When we come upon a process by which the same technologies are applied to different media, it seems logically unsound to assert that such a process is the result of "intelligence" only if the media have been manipulated by human hands.

We consider ourselves the only intelligent passengers on Spaceship Earth. In the next century, however, we may be in for a surprise. We may discover that there is some sort of captain on our spaceship that we have been too ignorant to recognize. It could really be that simple.

The BRAIN:
SEAT *of the* QUESTING MIND

A s evolution progressed and life forms developed, the world became
a place of increasing complexity. With that complexity came the need
for life forms to develop ways of interacting with their environments
and other life. The size and complexity of brains grew to accommodate evolutionary growth.

It is fascinating that whereas the various sensory organs and locomotive
organs—such as fins, limbs, or insect eyes—have a variety of forms and are
made by a variety of specialized cell structures, the brains of all life forms are
basically built up of the same type of cells, the nerve cells (or neurons). The
jellyfish, a very primitive type of life, uses the same type of nerve cell as the
sophisticated human being.

What matters is the variety and sophistication of the *interactions* between the nerve cells, the "software program," to use a computer analogy. It
is in this area that evolution has been operative and where the culminating
mystery of life appears: the emergence of mind, the human capacity for abstract, creative thought. The instincts of the lower life forms represent what
the computer programmer would call "hard-wired" programs (programs that

do not change over generations) that regulate certain aspects of the survival routine for those forms and are genetically transmitted from generation to generation. Human beings, though, generate part of their own software programs, which are transmitted not genetically but by education—an evolutionary breakthrough on the most advanced level.

Because the individual nerve cell is the key element in building up the neural networks (brains) in all life forms, we will first examine at the molecular level the structure and functioning of that remarkably specialized cell.

The average human brain is made up of 1 trillion cells, of which 100 billion are nerve cells that may connect with each other in 100 trillion ways. The 100 trillion atoms in a single cell are electrically bound together in various geometric systems such as the cell membrane, cell nucleus, DNA, mitochondria, ribosomes, Golgi apparatus, endoplasmic reticulum, and microtubules. Those are all internal structures for the life support of any of our cells.

In addition, special systems have been created to execute certain functions of the specialized cell. In the case of the nerve cell, its overriding function is to transmit signals, or "messages." We are far from understanding the full range of signals transmitted, but we are beginning to discern the details of the mechanism for signal transmission at the atomic-molecular level. The mechanisms at the molecular level display an astounding beauty and elegance.

The geometric shape of the nerve cell membrane has been severely distorted to serve as a signal transmission link (see Figure 6.1). On one side there are very short, branched, treelike structures called dendrites. On the surface of each branch of a dendrite, there are indentations that can be attached to the end of the single signal fiber, the axon from another nerve cell. The axon is the transmission cable stemming out of a nerve cell; its end point onto the dendrite branch of another nerve cell is called a synapse or synaptic button. The length of the axon is highly variable. In the human spine, the axon of a single nerve cell may reach a length of 3 feet.

Although the axon is a single transmission cable growing out of the cell membrane, it branches out into numerous synaptic buttons at its opposite end point so that the signal can be transmitted to a number of other nerve cells. At both the dendrite and the axon end points, a single nerve cell is connected to multiple other nerve cells. A single neuron receives signals from up to 10,000 other nerve cells.

In the modern computer, information in the form of data and built-in computer programs is processed through a number of individual transistor gates. These individual gates have only two possible responses: "open" or "closed" (i.e., "on" or "off"), indicating whether to let an electrical signal from the input side of the gate continue onward from the output side of the gate. However, it is the *assembly* of the transistor gates that does the *processing* of the information coming into the computer. This processing operation depends on a large number of "on" or "off" responses by the individual transistor gates.

The interior of the receiving nerve cell processes signals in a manner that is still largely unknown at the molecular level. But on the basis of that processing, the neuron sends a signal through its axon to the dendrite branches of other nerve cells. Thus, the individual nerve cell seems to operate like a computer, and because of the multiple hookups with other nerve cells through its dendrites, it resembles a computer that can send and receive information on a number of levels. A human brain, then, is the collective linking of 100 billion microcomputers. The analogy is not entirely correct, but it's important to realize that a single nerve cell is not just an on-off gate or switch, that it does quite a bit of data processing on its own.

All neurons share the basic functions of signal transmitter and data processor, but the diversity of dendrite branch structures can vary greatly in different areas of the brain.

The mechanisms for signal transmission are twofold and in a way are intertwined: an *electrical* signal transmission along the axon initiates a *chemica* signal transmission mechanism that acts across the synaptic button onto the dendrite membrane of another nerve cell (see Figure 6.1). That chemical mechanism is a relatively late development in evolution. Most synapses in the brains of mammals are chemical, whereas in many invertebrates the electrical synapse prevails.

NEURON
SIGNAL
TRANSMISSION

Figure 6.1

The membrane of the neuron cell is severely deformed into one signal cable out from the cell —the axon. This cable splits into several components that connect to a number of branch arms (dendrites and their synapses) in other nerve cells. When neuron A receives a number of signals from other nerve cells, it decides (by an as yet unknown decision-making process) whether or not to trigger an electric current pulse out along its axon. If it does send such a pulse, that signal travels at a speed of several feet per second to the dendritic synapse, where it triggers the release of a chemical signal in the form of special neurotransmitter molecules, which travel across the synaptic interface to the next neuron. This neuron in turn continues the signal activity.

Whereas in earlier life forms communications between neurons were purely electrical, more highly evolved life forms developed a combination of electrical and chemical signal transmission. This development permitted a response to build up gradually, depending on the amount and type of chemicals released.

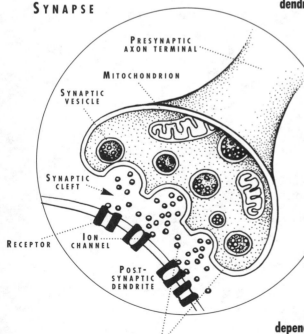

SYNAPSE

PRESYNAPTIC
AXON TERMINAL

MITOCHONDRION

SYNAPTIC
VESICLE

SYNAPTIC
CLEFT

RECEPTOR

ION
CHANNEL

POST-
SYNAPTIC
DENDRITE

NEUROTRANSMITTERS

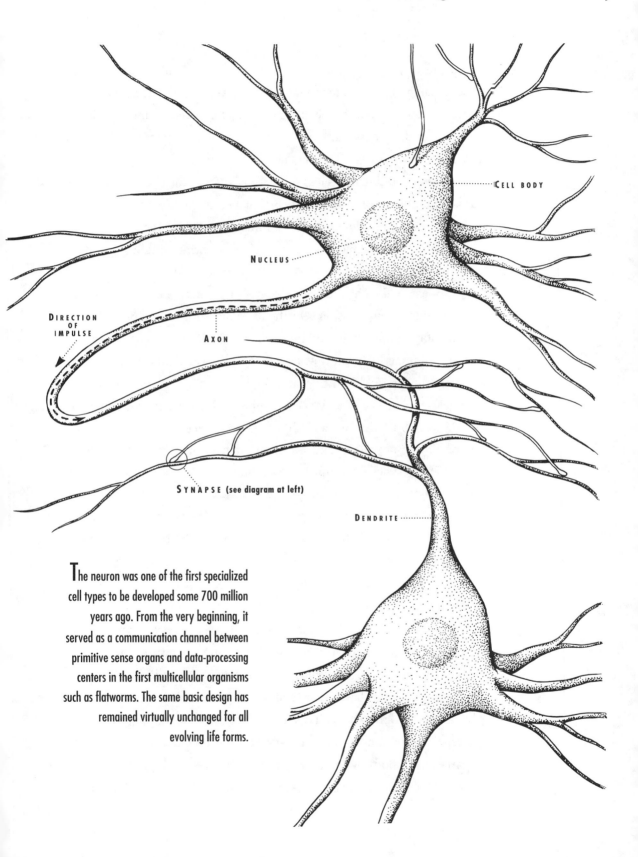

CELL BODY

NUCLEUS

DIRECTION
OF
IMPULSE

AXON

SYNAPSE (see diagram at left)

DENDRITE

The neuron was one of the first specialized cell types to be developed some 700 million years ago. From the very beginning, it served as a communication channel between primitive sense organs and data-processing centers in the first multicellular organisms such as flatworms. The same basic design has remained virtually unchanged for all evolving life forms.

An electrical synaptic impulse is like an axon nerve impulse; it is on or off. Once the signal impulse is set in motion along the axon, it continues to the very end, similar to the on-off response of digital computers. The chemical synapse, on the other hand, permits a gradual buildup to an "on" response. The signals of many chemical synapses from many nerve cells can lead to an "on" response from a single nerve cell that receives those signals. Thus, the chemical synapse provides a fine-tuning of a nerve network. It permits a sort of "almost" response—"almost on" or "almost off" (today known as "fuzzy" logic).

A scrutiny of the chemical synapse mechanism brings out astoundingly complex and ingenious details at the molecular level. The reason for the gradual response of the chemical synapse lies in the nature of its chemically conditioned signal transmission. On the axon side of the synaptic button, the cell membrane folds like a bud (see Figure 6.1, lower left side.) Inside this bud lie a number of chemical-carrying sacs, the vesicles. When the electrical signal impulse reaches the synaptic end of the axon, it triggers motion of the vesicles toward the synaptic cell membrane. The vesicles penetrate the membrane, which borders a fluid-filled synaptic space between the donor membrane of one nerve cell and the receptor membrane of the receiving nerve cell. The vesicles open up their sacs and release a rain of messenger molecules, maybe several thousand of them.

Those neurotransmitting molecules then attach to certain receptor molecules on the receptor membrane. Both types of molecules have been shaped geometrically and electrically to fit each other, much like enzymes have special geometrical shapes to link with the two participant molecules that will be catalyzed by the enzyme.

The receptor molecules are proteins that penetrate the cell membranes, which are made up of lipid molecules. The neurotransmitting molecules were originally thought to be of only three or four different kinds, but now over fifty different types have been cataloged. Many nerve cells do not necessarily activate other nerve cells; such activation happens mostly in the brain. From the brain, however, nerve connections go out to numerous different parts of the body. Many of the nerve cells eventually hook up with muscle cells, for example. This connection of nerve cells to other body cells constitutes the *peripheral* nervous system, as opposed to the *central* nervous system of the brain, which connects nerve cells to other nerve cells.

In the peripheral nervous system, the seemingly simple gesture of raising a hand is a result of the synaptic buttons of the nerve cells being triggered to action. In that case, the receptor molecules in the dendrite of the muscle cell will receive a special neurotransmitter molecule, which triggers a contraction or extension of the muscle cell depending on the particular brain action that is instructing the muscles to perform.

Another beautiful example of the special function of neurotransmitter molecules becomes evident when a human baby is about to be born. The pituitary gland in the mother's brain releases a special chemical, the hormone oxytocin. That particular chemical will travel in the bloodstream to different parts of the body and will enter numerous nerve cells. In most cells nothing will happen as a result. Only in the nerve cells attached to those muscle cells responsible for uterine contractions will oxytocin find the right receptor molecules to initiate contractions.

The ingenious design of the nerve cell and its signal transmission mechanism was laid down about 600 million years ago in the nervous systems of such creatures as the jellyfish. The evolution that has taken place since then has mostly involved the web of interconnectivity of the nerve cells. That focus ostensibly gave chance the task of developing meaningful "computer programs" in the brain network to further the survival and development of various species. In other words, the brains of the individual species needed to develop certain *instincts*.

One example is the elaborate building techniques and communication instincts that bees have for locating food sources. Those functions are precise, demanding programs that have been wired into the bee's brain, though that brain contains no more than 7,000 nerve cells. The complexity of such instinct programs boggles the mind, casting doubt on how they could have been developed by chance.

Let us consider another example. Lyall Watson, in his book *Lifetide,* describes an instinctive ritual involving a special wasp-tarantula species interaction:

This story involves the giant wasp, *Pepsis marginata,* which feeds its young larval offspring with giant tarantulas, *Cyrtopholis portoricae.* The wasp itself is vegetarian, but the larvae are carnivorous. The female begins to search for a tarantula when she feels an egg in her ovary is almost ready to be laid. Having found the tarantula, she begins to feel the tarantula out with her antennae to be sure it is the right insect food for her larvae when they are born. Normally the tarantula, upon being touched, would respond by attacking the intruder, but in this case nothing happens. The tarantula just stands there passively while the wasp feels it out. Having ascertained that it is the right insect for her larvae, the wasp begins to dig a hole 10 inches deep and slightly wider than the spider. Usually the spider will not flee the scene but will remain waiting. The wasp returns and slides on her back under the tarantula, feeling out the details of the spider's body. Finally, when she is in the right position and has found the special opening (which is where the legs join the body) in the hard outer shell of the spider, she puts her stinger into this opening. The sting has to go in with surgical precision at the right angle and the right depth so that the wasp finds the right nerve center that will paralyze the spider and render it motionless without killing it. During this entire operation, which can last for minutes, the tarantula makes no move to save itself.

Once the spider is paralyzed, the wasp drags it down into the tunnel she has dug; there she "handcuffs" each of the eight legs of the spider so that it cannot move out of the tunnel if it awakens. Then she lays a single egg beside this live food supply. The egg hatches a larva that then begins to consume the tarantula in such a way that the food supply stays fresh as long as possible—the vital organs of the spider are eaten last.

As Watson points out, the gradual development of this instinct by chance and natural selection is close to impossible, because this deadly game is an all-or-nothing situation. If the instinct is not completely developed at the beginning, the species will become extinct: either the tarantula will kill the wasp, so that the wasp will never learn to find the exact location of the spider's nerve center, or the larvae will die for lack of a food supply. Even Darwin had great difficulties explaining the evolution of the life cycles of certain insects "in which we cannot see how an instinct could possibly have originated" and "in which no intermediate gradations are known to exist" (quoted in Watson 1980, p. 166).

An interconnecting web of neurons started forming long ago, when the development of multicellular life forms began. Today we see relics of those early neural developments in flatworms and earthworms. Those primitive life forms have the rudiments of a brain in a head that may contain a few hundred nerve cells. As mentioned earlier, the growth in the complexity and functioning of the sensory organs proceeded in parallel with that of the sensory data-processing system—the brain. However, for invertebrates such as insects, their hard, chitinous bodies placed limits on the growth of their brains. Bees have no more than 7,000 nerve cells organized in nerve masses called ganglia, whereas mammals may have brains with several hundred million nerve cells. Humans have the most brain cells, 100 billion of them.

The rise of the mammalian brain is among the most fascinating aspects of the brain's evolution, so let us follow its line of development.

The first vertebrates (animals with backbones) were the vertebrate fishes, which had well-developed eyes and olfactory organs for smell and taste. The fish brain begins with a swelling at the end of the spinal nerve chord. That swelling is called the medulla oblongata. There, large nerve bundles fork out to different specialized data-processing centers such as the cerebellum, which processes the data produced by the fish's balance organs and swimming muscles. The optical lobes are highly developed to process visual data from the eyes of the fish. In addition, smell and taste data are processed in the relatively large olfactory globes. The cerebrum, the "thinking part" of the human brain, exists only in embryonic form in the fish. Yet modern fish, like sharks, do appear to be able to learn to a limited degree. Much is still unknown about the operation of the fish brain, but basically the fish do not "think" and do not possess intelligence as we know it. (It is worth noting that dolphins and whales are not fish but rather mammals that returned to the sea.)

Still, the fish brain is an impressive evolutionary development that must have taken place over not much more than 50 million years, concurrent with the development of the first eye lenses, olfactory organs, and swimming organs (rudimentary fins). Fifty million years is a very short time for chance to create such a coordinated development of several complex organs.

As the amphibians began to evolve into life forms adapted for life on land, the brain functions evolved in response to the more complex forms of locomotion, balance, food gathering, and escaping from predators. The reptilian ancestors of the mammals had a fishlike but considerably enlarged three-compartment brain for processing balance and muscle coordination. Some simple instincts surely were developed to promote survival, such as aggression, sex, and searching for food. The fight-or-flight response was also created in the reptilian brain. A new development in that brain was the creation of a center for coordinating smell and vision.

It is not very likely that the reptiles were thinking animals despite their more developed brains. The brain of the largest existing reptile of the dinosaur branch—the supersaurus—was only the size of an orange and weighed maybe half a pound. The nerve cells in that relatively small brain were probably occupied entirely with the instinctive tasks of moving, feeding, and protecting the huge 100-ton bulk of a body. (In contrast, the largest living animal today, the humpback whale, also weighs 100 tons but has a 1^{1}/2-foot-long brain that weighs 20 pounds.)

It is in our small mammalian ancestors that the next major brain development took place about 200 million years ago. The reptilian-amphibian line, which led to the development of the mammals, had shrunk in size to small mouse-sized nocturnal animals under the threatening dominance of the other branch of reptiles, the dinosaurs. The small mammals retreated into an "ecological niche," as the Darwinians would call it. That niche was characterized by small size and nocturnal activity. Eventually these creatures evolved into larger mammalian life forms once the dinosaurs became extinct 65 million years ago. But during that hiding period of about 100 million years, very important developments took place in the mammals' brains. Possibly as a result of their night forays, those little warmblooded animals had to rely greatly on their sense of smell; the smelling part of the brain began to grow in the direction of its adjacent part, the embryonic cerebrum. That change took place in order to process the increasing amount of smell data and to coordinate it with information from the other senses.

There were also new functions that the mammalian brain needed to become involved in, such as controlling body temperature independent of the external temperature and providing parental care for offspring over prolonged periods of time. Such features required an expanded brain capacity. In addition, with the disappearance of the dinosaurs, the mammalian ancestors could begin to operate in the daytime. Thus, the expanding vision-processing centers in the brain were also placed in the expanding cerebral cortex (the outer part of the cerebrum). It is the cerebrum that has grown explosively in the last few million years, making the mammalian brain the fastest-developing organ in evolutionary history. During the last 4 million years, the protohuman brain has tripled in size.

Among the now-living mammals there is a great variety in brain size relative to body weight. A rabbit, for example, if enlarged to the size of a person, would have a brain weighing 1 ounce, whereas a person's brain weighs fifty times more. A highly intelligent animal, such as the wolf, has a brain ten times smaller than a human's relative to body weight. A monkey's brain size relative to body weight is only five times less than the human's, and with the larger primates, the apes, that number shrinks to three or four.

Only humans show such a high brain-to-body weight ratio. The apparent reason for the remarkable growth in the cerebral hemispheres is that human beings' ancestors were tree dwellers. Some 65 million years ago, a branch of the mammals took to the trees, a move that had an enormous influence on the development of the brain. A life in the trees put powerful demands on the brain's coordination of eyes and limbs. So during the next 20 million years, the placement of the eyes in the head gradually moved forward to provide stereoscopic binocular vision; at about the same time, over 30 million years, the claw-equipped limbs by which the mammalian ancestors had made their way up into the trees transformed into supple, flexible hands. Thus, the first true monkey appeared about 35 million years ago.

What made the first monkey ancestors take to the trees is anybody's guess. Within the Darwinian conceptual framework, trees provided another ecological niche. However, after some 25 million years in the trees, some of the tree dwellers returned to the ground and became the immediate ancestors of the large primates: the gorillas, chimpanzees, orangutans, and human beings. Those animals adopted bipedalism, a walking mode on their hind legs, and they used their hands for crude toolmaking. The last part of the brain's development began about 4 million years ago, when the brain of *Australopithecus afarensis* was about 400 cubic centimeters in volume. In 2 million years it grew to the brain size of *Homo habilis,* and then in the last 2 million years the cerebrum grew explosively and doubled to its present size and weight of 3 pounds. In the last million years, 1 pound of grey matter has been added to the brain.

Let us briefly look at the modern human brain in the light of this sketch of its evolutionary history. The human brain is a unique organ in that it is, in evolutionary terms, a jumble of things. It is not a streamlined tool like the human hand, eye, or ear but rather a patchwork of grafts. The fish brain and reptilian brain are grafted onto the spinal cord; the limbic mammalian ancestral brain is grafted onto the reptilian brain; and, finally, the simian-human brain is grafted onto the two others (see Figure 6.2). Scientist and humanist Arthur Koestler, who in the last years of his life immersed himself in evolutionary problems, seriously thought that because of this patchwork the human experiment in evolution might fail. Yet as we examine the hardware and software of this patchwork, we see a beautiful ingenuity displayed in allocating different bodily functions and operations to different parts of the human brain.

Progressing directly from the spinal cord, we find a slight swelling called the medulla oblongata. This is the nerve center for such fundamental activities

EVOLUTION *of* *the* HUMAN BRAIN

Figure 6.2

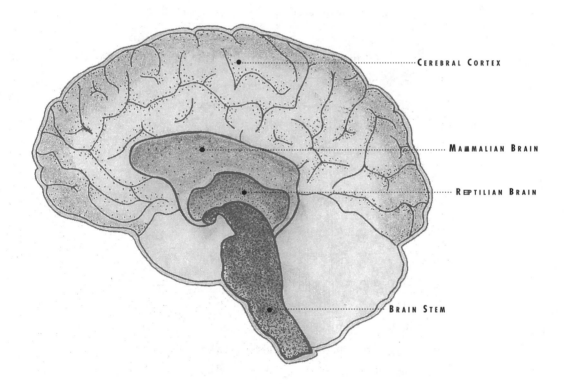

CEREBRAL CORTEX

MAMMALIAN BRAIN

REPTILIAN BRAIN

BRAIN STEM

The human brain consists of 1 trillion cells, of which 100 billion are neurons. The human brain is the culmination of at least 400 million years of evolution in three major stages. First came the reptilian brain (aggression, survival instincts) arising out of the brain stem. The mammalian brain (caring and nurturing emotions) was added some 65 million to 100 million years ago, and finally the large cerebral cortex (abstract mental thought) emerged only a few million years ago.

as breathing and the beating of the heart, or even swallowing and vomit-
ing—all very basic functions for keeping the body alive. The development of
the medulla oblongata goes back some 400 million years.

Above the medulla oblongata lies a bulbous protrusion of nerve cells
called the pons, which serves as a neural relay station between the spinal
cord and the mid-brain, where the thalamus and hypothalamus glands are
located.

Running through the medulla and the pons is the reticular formation,
with its reticular activation system. That system is the clearinghouse for all
the raw sensory data that come in via the spinal cord from the rest of the
body's cells. A vital function of the reticular formation is to act as a kind of
autopilot for the body's general muscular framework. If we wish to stand upright, a command goes from the cerebral cortex (the outer part of the cerebrum) to the reticular formation, which then executes the command to the
trillions of muscle cells involved in the command "stand upright." The system, in other words, is like a constantly buzzing telephone exchange where
innumerable nerve cell messages come in and a very rapid screening process
takes place. The screening determines which signals are important and novel
enough to be reported to the higher portions of the brain.

Immediately above the top of the brain stem lies a pair of ball-like structures the size of robins' eggs: the thalamus, which is the main relay station
for sensory input into the brain (except for the sense of smell). The thalamus
appears to be the center of awareness in reptiles. In mammals, the thalamus
sorts the sensory information and relays it to the proper specialized sensory
centers in the higher brain, the cerebral cortex, which is the mammals' center of awareness.

In front of the brain stem and directly under the thalamus lies the vital
organ called the hypothalamus. It is roughly the size of a pea and accounts
for only 0.3 percent of the whole brain weight, but it is considered to be *the*
central monitoring and control station for a large variety of bodily functions
(see Alvin and Virginia Silverstein, *World of the Brain*, 1986, pp. 60–92).
The hypothalamus receives manifold nerve messages, such as information
about blood pressure, heartbeat, body temperature, and danger signals.

Upon receiving that information, the hypothalamus sends messages to other organs to produce chemicals or perform muscular operations in order to cope with the situation. In addition, the hypothalamus has processing centers relating to hunger, satiety, and thirst. There are also centers to monitor sex, fear, anger, pleasure, and pain.

A cable of nerves connects the hypothalamus with the pituitary gland. Many of the decisions made by the hypothalamus on the basis of incoming sensory data are acted upon by chemical hormone secretions of the pituitary gland, which in turn regulates other glands, such as the sex glands.

The pineal gland, the rudimentary third eye (a relic from a stage in the development of fish), is still connected to the optical nerves that link the eyes to the higher brain. It is now believed that the pineal gland regulates the monthly and diurnal cycles of human beings.

Finally, on top of the brain stem are four little humps called the superior and inferior colliculi. Those humps are the remnants of the visual and acoustical data-processing centers in fish and reptiles. The superior colliculi are the remnants of the optical lobes, and the inferior colliculi are vestiges of the acoustical lobes. Both serve as a kind of early warning system. The former regulates eye blinking, pupil opening, and focusing the eye lens, whereas the other controls the ear's sensitivity and makes the body react to loud noises.

A pair of bean-pod-shaped structures, called the hippocampus, are nestled under the hypothalamus and seem to be instrumental for stimulating the formation of long-term memories in the higher, newer brain. Likewise, another pair of almond-shaped structures, the amygdala, connect with other parts of the mid-brain, as well as the seeing and hearing centers of the cerebral cortex. The amygdala seems to be involved in learning and remembering by providing an emotional motivation factor.

Taken together, the thalamus, hypothalamus, and other structures of the mid-brain are called the limbic system. Those parts of the human brain evolved in our mammalian ancestors during a time period from 300 million years ago until about 100 million years ago. The limbic brain provided the mammalian animals with a set of automatic reactions—instincts—to the information gathered by their senses. The limbic system may be likened to an array of large-capacity computers that run a set of software programs (instincts) to monitor and regulate all the major life functions of the body automatically.

Thus, an incredible amount of data input and data processing is handled by billions of nerve cells without us consciously being aware of it. This is an absolute necessity for not overloading the conscious parts of the brain: the cerebellum and the cerebral cortex. Those newest parts of the brain are like a top executive's office, where a number of secretaries have screened out visitors with unnecessary information; only visitors who rightfully should command the attention of the top executive are allowed to be there.

We have mentioned that all sensory input except that related to smell goes into the thalamus. Messages from the nose go directly into the cerebral cortex. That design feature is an evolutionary relic from the time of the early mammals more than 100 million years ago, when the cerebral cortex began to evolve out of the smelling part of the brain of the small, nocturnal mammalians. The analysis of smell signals and their coordination with sensory input from other organs apparently led to the emergence of a more sophisticated system of nerve cells that began to overlay the smelling part of the brain, as we see today in primitive mammals like the rat. In the rat the part of the brain that handles smell is overlaid by a smooth cloak of gelatinous nervous matter called the neopallium. In dogs that new cloak is greatly expanded and folded into innumerable wrinkles. The mindfulness of Nature's designs is revealed once again. Not only is the vulnerable gelatinous mass of the brain encased in hard, thick skull bones, but to optimize the brain's information-processing capacity, the surface tissue of the new brain that began to evolve some 100 million years ago developed crinkled folds. The result is somewhat like crumpling the large area of a newspaper page into a small ball with folds and crinkles. In that manner, more brain surface area could be packed into the volume of the skull. (In some sense, the folds are analogous to the highly economical packing of the human chromosomes— some 6 feet in length—into the tiny ball of a cell 0.1 millimeter in diameter.)

Lying on the back of the limbic brain is the cerebellum, the seat for the coordination of body movements. The cerebellum constantly monitors information received from the cerebral cortex concerning ordered body movements and compares the order with its actual execution. The deceptively simple operation of stretching the hand to touch a cup of coffee and bring it to the mouth for drinking involves a complex series of nervous interactions among the cerebral cortex, the cerebellum, and the muscles of the arm, hard, fingers, and eye. The cerebellum is also the center for learning complex muscle operations like skating, playing tennis, or driving an automobile.

We might ask ourselves where in the evolutionary life of an animal there emerges an individual generation having an individual, self-made "software program." Robert Jastrow, in his book *Enchanted Loom: Mind in the Universe,* equates that stage with the emergence of intelligence. To Jastrow, intelligence is the capacity for learning and the ability to adopt flexible strategies as a result of that learning process. Thus, intelligence emerged on Earth with the appearance of the mammals about 200 million years ago. The brain-to-body weight ratio of the early mammals, which were mouselike creatures, was five times that of the giant dinosaur *Tyrannosaurus rex* and twenty times that of the plant-eating dinosaurs. Much of the relative growth of the mammalian brain was marked by the early development of the cerebrum in the form of the neopallium and the cerebral cortex.

Today, the cerebrum and cerebellum constitute about 90 percent of human brain weight, most of that within the cerebrum. It is for the cerebral cortex that most of the scientific knowledge has been gathered, through mapping done by brain researchers. Yet there are those researchers who strongly believe that the "vast silent regions beneath the cortex may one day yield their own exciting surprises" (Silverstein and Silverstein, *World of the Brain*, p. 83).

The cerebrum is divided into two hemispheres, the left and the right. Each hemisphere is primarily involved in the activities of one half of the body. For reasons unknown, the nervous system connecting the two halves of the body crosses over in the cortex, so that the left half of the brain governs the right half of the body, and vice versa. There is also a vertical crossover, so that the nerves from the face and head are connected to the bottom of the cerebrum and the nerves of the feet and lower parts of the body are connected to the top of the cerebrum. In addition, the two brain hemispheres are connected by a thick nerve cable called the corpus callosum. That design provides for a continuous flow of information between the two hemispheres, enabling activities to be coordinated properly.

The emerging map of the cerebral cortex allocates specific regions for specific functions. For example, a motor strip and a sensory strip are found across the top part of the cerebrum and down along the sides. There, the command centers and information-processing centers are located for limb-motor activities, like lifting a finger, and for touch.

The front part of the cerebrum is composed of what are called the frontal lobes. These lobes apparently are the executive center for thoughts, plans, decisions, and various other aspects of personality. The back part of the cerebrum constitutes the occipital lobe, where the vision center is located. The side parts are called the temporal lobes, where the centers for speech and hearing are located. Finally, the top part of the cerebrum toward the back is composed of the parietal lobes, where the center for hand skills is located.

Thus, in the cerebral cortex are found specialized learning computers, which manage extraordinarily complex data processing. A whole book could be written about the visual recognition of familiar faces and patterns. Pattern recognition has been a highly touted development in software for the most recent manufactured computers. Speech represents another remarkably complex data-processing facility in the cerebral cortex. It is highly significant that in the infant brain of the human animal there exists a larger number of unwired blank circuits in the network of the brain cells than in any other animal. As the human animal learns from experience, software programs are developed in the brain, so that, for example, the human growing up in Spain will speak Spanish.

With the development of the human brain, we have moved into the area of Nature's grandest design scheme. In the creation of the human animal, 2 billion years of primitive cell development and 800 million years of multicellular life form development culminate—though with a spectacular brain development (the cerebral cortex) that took only a few million years.

In our perusal of the variegated parts of the reptilian, mammalian, and human brains, we have encountered time and again arrays of sophisticated software programs that monitor and coordinate a great variety of bodily functions. Some of these software programs are unique to the human species and can have developed only over the relatively brief period of 1 million years, when the last pound of gray matter was added to the protohuman brain.

This extra pound of organic brain matter is not simply another pound of muscle tissue. It is the growth of at least 30 billion new nerve cells, which provide for at least 30 trillion new neural connections. This new pound of nerve tissue has to be coordinated with the older part of the brain, and it provides a unique tool capable of creating civilization and culture, which define the essence of human beings and set the species apart from all others

And indeed chance—the Darwinian dice thrower—in this special case more so than any other in biological history, is up against formidable odds. A little simple arithmetic will make the point. In older times of human history, a life span of 40 years per generation is a reasonable assumption. Thus, during a period of 1 million years 25,000 generations would have existed. Early human beings are not thought by anthropologists to have been very numerous, so let us assume 1 million individuals per generation. This makes for a grand total of 25 billion people upon which chance can play its game during 1 million years.

Chance operates on the genetic blueprint of each human individual either by altering one genetic code letter in a gene (a point mutation) or by reshuffling gene segments (sexual recombination). No matter how the detailed arithmetic is worked out to arrive at the total number of genetic changes in these 25 billion people, the incontrovertible fact remains that over half of these people have lived on Earth during the last 4,000 years. Yet during these four millennia no fundamentally new brain faculties have been created, nor has any significant growth of brain volume taken place. *If chance was the creative agent in the distant past, why not in the last 4,000 years?*

The most difficult creative achievement in evolutionary history took place in an "instant"—a flash of generations. To have the gall to claim, as the Darwinists do, that this was the product of chance and natural selection is utterly nonsensical.

But there is more to it than this. The human brain is assuredly more than an assembly of self-generated computer programs. The frontal lobes of the new brain are deemed the seat of the individual personality. There resides the executive center for decision making, planning, and creative thought—what we would call in everyday language the mind. It may be significant that in the index of Robert Jastrow's *Enchanted Loom: Mind in the Universe,* there is no entry for the word *mind,* even though that word appears in the book's title. However, there are twelve entries for the word *intelligence,* which Jastrow equates with the capacity for learning.

Is the capacity for learning all there is to the mind? What about creativity and the ability to generate new knowledge for future generations to incorporate into their brains? There is a vital component missing in the natural scientist's definition of mind, and it is about time we enter a different conceptual Universe, that of the humanist. The humanists are the only ones who have dared to bridge the gap between mind and Nature. It would behoove us to examine their way of thinking to see whether it can bring us out of the present impasse between seeing Nature's extraordinary designs and the palpable impotence of chance and natural selection as explanatory principles for the creation of those designs.

Throughout this book, in our search for knowledge and understanding of the evolution of life on Earth, we have compared time and again the developments of particular organs, life forms, and life support systems like photosynthesis with the odds that chance and natural selection could have been the creative agents. Now that we have examined the brain, with its culmination in the development of the human brain over only 2 million to 3 million years, we see the Darwinian paradigm face its most formidable challenge: accounting for the appearance of human intelligence by a pit ful few throws of the dice.

Darwin professed to be an agnostic when it came to chance being the creative agent behind intelligence. Beyond all analogies between compu ers and intelligence lies the realm of the mind, which is more than intelligence and in the humanist's Universe is characterized by artistic creations, the re-ligious lives of the mystics, the great social innovations, and the world of human culture and civilization.

The last section of this book will be a search into the world of the hu-manists to see whether we can find aspects of the phenomenon of mind, as they see it, that might bring us out of this flagrant mismatch between the supposed creative power of chance and its actual impotence in mathemati-cal terms. We shall start this search by examining the origin of Darwin's work and the subsequent developments by Darwin's followers, the neo-Darwinists.

The DARWINIAN REVOLUTION

T he appearance in 1859 of *On the Origin of Species* by Charles Darwin represented a major watershed in human intellectual history, compa- rable in importance to the *Principia* by Sir Isaac Newton in 1687 and the "Special Theory of Relativity" by Albert Einstein in 1905.

Yet, whereas the treatises of Newton and Einstein abound in mathemat- ical equations, there is not a single line of mathematics in Darwin's work. The mathematical tools of Einstein and Newton led to theoretical predictions that could be confronted with observations for numerical verification. Even today, however, no such mathematical predictions exist in evolutionary bi- ology. As a result, as eminent evolutionist Theodosius Dobzhansky puts it, "Since we cannot predict evolution, the theory of evolution has sometimes been derided as only quasi-scientific" (Ayala and Dobzhansky 1974, p. 329).

Even as recently as 1974, noted philosopher Karl Popper said that "Dar- winism is not a testable scientific theory, but a metaphysical research pro- gram" (Mayr 1988, p. v). In a curious way this sentiment is echoed by Har- vard professor Ernst Mayr in his book *Toward a New Philosophy of Biology,* wherein he states, "Darwinism is not a simple theory that is either true or false, but is rather a highly complex research program that is being continuously modified and improved" (p. 535).

Indeed, although Darwinian evolutionary theory was of absolutely vital importance in establishing an evolutionary sequence of species, Darwinian theory and its further refinements have utterly failed to put forward a believable and acceptable *explanation* for this evolution.

This state of affairs should not be surprising. Evolutionary biology as an intellectual discipline is only a little more than 100 years old. It would be remarkable if the first explanatory scheme put forward by Darwin—chance and natural selection—should be the correct one. It appears much more probable that the Darwinian explanation of evolution would fall in the category of, say, the theory of epicycles put forward by Ptolemy in the second century A.D. to explain the zigzag motions of the planets. Although the Ptolemaic theory numerically accounted for the observations, it had no foundation in physical principles; it was merely a mathematical and geometrical exercise. The correct physics was developed by Newton through his mathematical formulation of the law of gravitation and its application to Kepler's laws of planetary motions. Darwinian evolutionary theory still awaits its "Newton." If we proceed within the confines of the conventional physical theory of our days, however, we may have to wait forever.

The lack of mathematization and of predictive capability even in current post-Darwinian theory is the reason the Darwinian revolution is discussed in this book in the context of humanism. Darwinian theory is not a bona fide hard natural science like physics, nor is it as soft a science as, for example, anthropology, although it leans more in that direction. We may see, however, that in the development of molecular biology and its coupling to something known as information theory, evolutionary biology may occupy a unique position as the only natural science deeply related to the humanistic disciplines. Thus, evolutionary biology may bridge the gap between the two fundamental areas of human intellectual endeavors: the creative arts and the natural sciences.

The indisputable virtue of *On the Origin of Species* was that it planted firmly and irrevocably on the intellectual map of the Western world the concept of evolution of the species from predecessors as revealed in the study of fossil records. So far we have dedicated several chapters to revealing the intricacy and beauty of that evolutionary past and to delineating the time scales involved.

As is usually the case in the history of the sciences, Darwin's ideas on evolution did not appear out of an intellectual vacuum. The biblical belief in a seven-day history of creation had already received its first shattering blows through the observations and writings of James Hutton (1726–1797), the father of modern geology, followed by *Principles of Geology* by Charles Lyell (1797–1875). In those writings geological time scales of *millions* of years were proposed convincingly for the first time to account for observations concerning erosion.

Lyell even continued work in biology in which the principle of natural selection was enunciated and used to explain why species that developed damaging variations did not survive. That work was followed by some papers by Edward Blyth ("The Varieties of Animals," published in 1835, and "Distinctions Between Man and Animals," in 1837, cited in Hoyle 1982, p. 29) wherein Blyth argued for the operation of natural selection on existing species but reserved the biblical special creation for the species' appearance on the biological scene.

Lyell and Blyth had themselves been preceded by French naturalist Jean Baptiste de Lamarck (1744–1829), who in his 1809 book *Philosophie Zoologique* stated, "May it not be possible . . . that the fossils in question belonged to species still existing, but which have changed since that time and have been converted into the similar species we now actually find?" Lamarck then proceeded to speculate about the causes for that evolution and launched the much maligned Lamarckian theory of evolution. According to that theory, biological characteristics acquired during the life of a generation of a species could be reflected in that generation's genes. Those characteristics would then be transmitted to following generations.

From those writings it should be clear that Charles Darwin's thinking on evolution and evolutionary biology owed great intellectual debts to many predecessors. Even some of Darwin's contemporaries were influential. One such contemporary was British naturalist Alfred Russel Wallace (1823–1913), who worked in faraway places such as the Brazilian rain forest and the Indonesian jungles. While in Indonesia, Wallace sent off a manuscript to Darwin for a paper titled "On the Tendency of Varieties to Depart Indefinitely from the Original Type." In that paper, Wallace in effect anticipates the Darwinian theory of natural selection in evolution. That paper evidently catalyzed Darwin into first putting together a paper of his own, which was read before Wallace's paper at the meeting of the Linnaean Society on July 1, 1858, and then to completing with fervor his manuscript for *On the Origin of Species.*

In that book Darwin draws on his numerous observations from a five-year voyage (1831–1836) on the ship *Beagle* during his twenties. Darwin, who was born in 1809 and died in 1882, weaves that material into an observational buttressing of his theory of the origin of the species. The theory, according to well-known modern naturalist Ernst Mayr, consists of four principal postulates running through a broad explanatory scheme. (See that review by Mayr in *The Fossil Record and Evolution,* p. 8.)

The first two postulates, Mayr notes, are consistent with earlier Lamarckian philosophy. Darwin postulated that evolution does take place as evidenced by the fossil records. Darwin's study of those records convinced him that some species became extinct while new species appeared and flourished for certain periods of time. His second Lamarckian postulate was that the process was gradual and continuous, with no sudden leaps or changes.

In the third postulate, however, Darwin and Lamarck part ways in a significant manner. Darwin was convinced that similar organisms descended from a common ancestor, whereas Lamarck assumed that each species arose by spontaneous creation and, independently of other species, strove to attain perfection in design. Thus, in Darwin's view, all the insects evolved from a common ancestor, as did all mammals, including humans. That hypothesis—humanity sharing a common ancestor with the apes—offers testimony to Darwin's intellectual courage in the face of the Victorian religious climate of his time, and for those views he was relentlessly attacked and ridiculed.

Darwin's fourth postulate was that evolutionary change occurred as a result of natural selection. That process, he argued, was twofold. The first step was the production of variation in a given generation of a species. His own vast observational material from his journey on the *Beagle* was drawn upon—for example, the variation in finches he observed on the Galapagos Islands. Seeing the variations in beak sizes and shapes, he noted that "in one small intimately related group of birds, one might really fancy that . . . one species had been taken and modified for different ends." The cause for that variation he did not understand, and his observations preceded the important papers on genetics published by Austrian monk Gregor Mendel (1822–1884) in 1866.

However, Darwin formulated the second step in the operation of natural selection after reading the works of the father of population dynamics, Thomas Malthus. The celebrated sentence by Malthus, "It may be safely pronounced therefore, that population when untouched, goes on doubling itself every twenty-five years, or increases in geometrical ratio" (Malthus 1798), must have crossed Darwin's mind. Malthus also linked that explosive population growth to the availability of environmental resources for nourishing such a growing population, especially for competing populations. According to Malthus, such growth would lead to an inevitable conflict between those rapidly growing populations regarding their dominance over the available natural resources.

So arose Darwin's concept of the regulatory and influential mechanism of natural selection, which has been popularized under the phrase "survival of the fittest." As Ernst Mayr points out in his penetrating book *The Growth of Biological Thought,* even here Darwin was not an original thinker, nor for that matter was Malthus. Mayr traces the concepts partly to Benjamin Franklin and to British philosopher William Paley (1743–1805). Paley was one of Darwin's favorite authors, despite Paley's declaration that "there cannot be a design without a Designer."

Darwin himself was well aware of the difficulties his theory of natural selection would have in explaining the details of exquisite sensory tools like the eye (see Chapter 5). However, Mayr displays a much less hesitant attitude about the same problem. He writes, in "Origins of Evolutionary Novelties" (*Toward a New Philosophy of Biology*):

> The most striking case, of course is the evolution of the eye. As Darwin already suggested, nothing is needed for their evolution but the existence of light-sensitive cells at the surface of an organism to initiate a chain of structural and functional events, culminating in a more or less highly developed photosensitive organ. There is no need to postulate rare or unique events, since eyes evolved in the animal kingdom at least 40 times independently.

This statement gives a glimpse of some scientists' tendencies to leapfrog over the difficulties of explaining all the *micro*-level events that must have taken place to lead from the light-sensitive cell of a worm to the superb optical instrument of the animal eye. A far greater amount of high-tech information has been put into the genes of the animal than into the genes of the worm. The ultimate question is: *How has this information been put into the genes*? Did it occur by chance and natural selection, as the Darwinists will have it, or by some other means?

Recently some remarkable computer simulations on the evolution of the vertebrate eye have been carried out by Dan Nilsson and Susanne Pelger (1994) at the University of Lund in Sweden. Their simulation shows that a kind of eye lens geometry will form from a flat patch of light-sensitive "tissue" in the course of only 400,000 years under reasonable assumptions of random tissue deformation and refractive index variations.

Their work was enthusiastically commented upon in an article in *Nature* by Richard Dawkins (1994, pp. 6901–6902), author of *The Blind Watchmaker,* a passionate defense of classical neo-Darwinism. Like Mayr, Dawkins is a macrobiologist, but he ventures into a very simple computer simulation in his book to "prove" that accumulated small steps of chance variations can lead to meaningful results. He tells of a computer program he created that generates as a first step an arbitrary sequence of twenty-eight letters to approximate a given sentence: "methinks it is like a weasel" (*The Blind Watchmaker,* pp. 47–48). He then announces that his computer program was able to retrieve the original sentence in only forty-one trials, or "generations," and that it took 11 seconds. But this is an intellectual sleight of hand of the worst kind, because the whole point is that the computer should not know beforehand the content of the model sentence and be able to compare each successive try with this model sentence. Dawkins's book has passionately argued that evolution is a *blind* watchmaker.

The weakness in all computer simulations is that they in no way take into account the detailed cellular biophysics at the molecular level in their definitions of computer variables. In Nilsson and Pelger's computer simulation of the eye, for example, it is obvious that a "sharper vision" criterion will encourage both tissue deformation in the direction of circular symmetry and a functional lens geometry (albeit a single lens, not a doublet as in the case of the trilobite eye). However, it should be stressed that the computer "tissue" is only a mathematically idealized plane surface. How can such tissue deformations be achieved at the cellular level, at the protein-manufacturing level, and at the DNA level in order to preserve a small design improvement for future generations? The chemical complexity of the interior of even a single cell is staggering. How can these complex substructures be integrated into the evolution of the eye? Anyone who has studied the remarkable integration of a variety of advanced technological features of the eye—from the chemical photomultiplier structures of rods and cones to the variable-focus deformation of the eye lens—should realize that mere tissue deformations and refractive index variations are not enough to explain the technical sophistication that goes into the design of the vertebrate eye.

In some measure, qualitative intellectual handwaving has permeated all thinking in evolutionary biology since the appearance of the Darwinian bible. And sadly enough, but as so often happens with the disciples and converts of a great master, the original thinking has become dogmatized and codified.

Darwin himself, toward the end of his life, adopted a kind of agnostic attitude toward religion and life. He wrote, "My theology is a simple muddle. . . . I cannot look at the Universe as the result of blind chance, yet I see no evidence of beneficent design in the details" (quoted in Jastrow 1981, p. 100). There is no such modest hesitation in modern evolutionist George

Gaylord Simpson, who categorically declares that "evolution achieves the aspect of purpose without the intervention of a purposer, and has produced a vast plan without the action of a planner" (Simpson 1947). As we shall see, the modern development of neo-Darwinian theory still produces such categoric statements without a shred of hard mathematical analysis. Indeed, the past and present states of evolutionary theory bear much more resemblance to religious faith than to hard-nosed natural science.

Of course, anyone who tries to grapple with a fundamental critique of Darwinian theory had better read up on the thinking of modern Darwinists. Harvard professor of zoology Stephen Jay Gould (1992) had this to say in a critical review of *Darwin on Trial,* a 1991 book by UCLA law professor Phillip E. Johnson:

> Now I most emphatically do not claim that a lawyer should not poke his nose into our domain; nor do I hold that an attorney couldn't write a good book about evolution. . . . But to be useful in this way, a lawyer would have to understand and use our norms and rules, or at least tell us where we err in our procedures. . . . Johnson continues to castigate evolution for old and acknowledged errors. . . . He quotes Ernst Mayr from 1963, denying neutrality of genes in principle. But much has changed in 30 years and Mayr is as active as ever at age 87. Why not ask him what he thinks now?

So what are the norms, rules, and procedures of modern Darwinists? Anyone trying to learn where modern Darwinism stands would do well to turn to Ernst Mayr's recent books. In 1988 Mayr published *Toward a New Philosophy of Biology,* then in 1991 *One Long Argument.* Those two books leave one with an impression that considerable controversies exist over details among Darwinists. Mayr does not shrink from exposing those differences of opinion in his books. One reason for such differences lies in the diversity of professions that fills the spectrum of the Darwinists' camp.

Broadly speaking, there are two major factions: the microevolutionists, represented by the geneticists and the molecular biologists, and the macroevolutionists, represented by the naturalists (such as botanists and zoologists) and the paleontologists. The microevolutionists study the genetic makeup of the germ cells (eggs and sperm) of a given species, called the *genotype,* whereas the macroevolutionists study the fully grown individual, the *phenotype,* of a given species. In other words, the microevolutionists study the details of the genetic blueprint as evinced in the cellular chromosomal DNA. The macroevolutionists study the appearance and function of the large-scale traits (such as beaks, furs, and eyes) of the whole organism built from that genetic blueprint.

Mayr paints a vivid picture of the battle that occurred between the two groups in the 1930s and 1940s in a chapter of *Toward a New Philosophy of Biology* entitled "Does Microevolution Explain Macroevolution?" Ultimately a synthesis of the two points of view was achieved under the name neo-*Darwinism*. This synthesis was carried out by zoologist Ernst Mayr himself, geneticist Theodosius Dobzhansky, paleontologist George Gaylord Simpson, biologist Julian Huxley, and geneticist G. Ledyard Stebbins and others.

The controversy centered on whether the Darwinian principles of chance and natural selection acted on the genetic blueprint directly, on individuals in a population (large group) of a given species, or on both levels. Although Mayr states that "now, 50 years later, the controversy still seems undecided," he goes on to state several formulations that he evidently considers settled. Among these are the following:

1.) The individual organism is the principal target of *selection.*
2.) Genetic variation is *largely* a chance phenomenon. Indeed stochastic (random) processes play a large part in evolution.
3.) The genotypic variation that is exposed to selection is primarily a product of (sexual) recombination and only ultimately of *mutation.*
4.) All speciation (i.e., creation of a new species) is simultaneously a genetic and populational phenomenon. (*Toward a New Philosophy of Biology,* p. 532)

These basic formulations, written in 1988, should be representative of current thinking in a major segment of the neo-Darwinist camp. In examining these formulations more closely, we discern a mixture of two levels of organization: the microbiological one of the genetic blueprint (the genotype) and the macrobiological one of the full-grown organism (the phenotype) as a member of a population.

From these four "credo" statements it appears that, according to the neo-Darwinists, natural selection operates only on the individual organism (that is, the phenotype). In addition, the fully grown organism provides a large number of offspring that are slightly different genetically and thus will respond slightly differently to the environment and the concurrent selection pressure.

Within a given population, genetic variations (that is, variations in the genetic blueprints) are brought about by sexual recombination (a mixing of the father's and the mother's genes). This sexual recombination happens by chance and is the prevalent means of gene variations—rarely do mutations (or genetic "letter" changes) occur individually.

Population genetics, the study of the genetic variation in a large number of individuals pertaining to the same species, was developed in the 1920s and 1930s by mathematicians such as R. A. Fisher and Sewall Wright, later followed by geneticists M. Kimura, C. C. Lin, and others. The population geneticists employ a wide and sophisticated array of mathematical techniques. However, it appears these mathematical tools have been applied only to explaining secondary characteristics within a single species (for example, longer beaks or variation in coloration). These techniques have never been applied satisfactorily to explain the appearance of new species with genuine genetic innovations.

It is highly significant that Mayr's most profound explanatory difficulty as a macroevolutionist is to account for the "origin of species"—the very phrase in the title of Darwin's book. In the neo-Darwinian "credo" listed above, the fourth point concerns the appearance of a new species. Such speciation is attributed simultaneously to genetic and populational factors. Mayr, in an interview in the July 1994 issue of *Scientific American,* still champions the role of the population factor in the process of speciation. In 1954 he advanced the hypothesis that speciation is favored to occur in small founding populations that are geographically isolated from the rest of their kind (see pp. 157–180 in Huxley, *Evolution as a Process*). However, Mayr shows he understands that sexual recombination in a given population *annot be the source of true genetic innovations. He writes, "The mixin₃ of genetic factors of both parents can supply an almost unlimited supply o ge-netically new individuals in every generation, but this process consists ɔnly in the intermingling of already existing variations. *The origin of entirely new genetic factors remains unexplained*" (*Toward a New Philosophy of Biolo₃y*, p. 511, italics added).

Mayr also devotes four essays in *Toward a New Philosophy of Biolo_y* to the question of species and speciation. These essays give the distinc⊤ im-pression that there is much confusion and controversy, especially amon₃ the macroevolutionists: "Perhaps it is a vain hope to expect an ultimate re₅olu-tion to the species problem, because the species play such different ro₃s in the thinking of various kinds of biologists" (p. 313).

Nevertheless, Mayr acknowledges that there exists a basic genetic cㄴange underlying the phenomenon of speciation. But for the new species tc suc-ceed, its members must face the adversity of the environment, with its pres-sure of natural selection. The problem of how the detailed change in tㄴe ge-netic structure takes place is left unanswered. Mayr speaks of "our igncⴰance of the genetics of speciation" (p. 362). He appears to embrace the ide◢ that speciation involves drastic restructuring of gene complexes rather than point mutations in individual genes (pp. 362, 471).

In order to penetrate deeper into the problematics of speciatiㄱ, we must turn to the microbiologists and listen to their statements on this issue.

In the authoritative undergraduate-graduate text *Molecular Biology of the Cell* (Alberts et al. 1989), we find the point of view of microbiologists, ⴰn sev-eral respects different from the "holistic" one put forward by Mayr anc other macrobiologists. James D. Watson, co-discoverer with Francis Crick of the DNA helix structure, is a co-author of that book. The opening chapter states:

> Consider the process by which a new anatomical featurᴇ—say an elongated beak—appears in the course of evolution. A rⴰndom mutation occurs that changes the amino-acid sequence of a ɔrotein or the timing of its synthesis and hence its biological activi y. This

alteration may by chance affect the cells for the formation of the beak in such a way that they have one that is longer. But the mutation must also be compatible with the development of the rest of the organism: only then will it be propagated by natural selection. (P. 31)

Clear enough, this is a typical Darwinian belief in chance mutations at the genetic level, the survival effects of which are "sieved" by natural selection.

Because it is an incontrovertible fact that the grown organism arises from its own genetic blueprint, it would appear logically consistent to follow the microbiologists' genetic point of view to analyze the problem of speciation. Thus, a complete analysis of the genome (the DNA chromosome complex) of one species should be contrasted with a similar analysis of the genome of a "nearby" located species on the tree of life. What are the differences in the enzymes and other proteins the two species use? What are the differences in the numbers of genes, the types of genes, their interactions, and so forth?

Eventually we would come down to a question that was posed by G. L. Stebbins and F. J. Ayala (*Scientific American*, July 1985, paraphrased here): How have the *differences* in meaningful sentences of DNA accumulated in the course of these species' evolution? These differences in meaning (that is, in information content), whether in specific genes or the architecture of the gene complex, must be the crucial factor accounting for the creation of new species.

However, it is quite apparent from *Molecular Biology of the Cell* that Mayr's holistic concept of the DNA genome complex is shared to some extent by the microbiologists. There are domains of the DNA structure that seem to control the triggering of subsequent gene interactions. One such domain is the homeobox, a group of genes whose function is to divide the early embryo into a band of cells having the potential to become specific organs or tissues. Evidently a point mutation in the homeobox could have drastic consequences for the development of the organism, beyond changing just one letter in the DNA sequence. Thus, additional information beyond that of the genes is built into the interaction patterns of the genes, that is, into the DNA *architecture*.

The microbiologists also subscribe to the same view as the macroevolutionists concerning the value of sexual recombination in furthering and accelerating the development of the DNA genome of a given species (Alberts et al. 1989, p. 599). However, as both macro- and microevolutionists admit, "The evolution of complex organisms involves more than simply introducing improved forms of existing genes: it means creating new genes to serve new functions. How does this occur?" (Alberts et al. 1989, p. 841).

The answer, the microbiologists believe, lies in the duplicate chromosome set that most life forms carry, one set coming from each parent The microbiologists suggest that "new genes can be created by the mutation of the spare copies in existing genes" (Alberts et al. 1989, p. 843). These new genes, by sexual recombination, are then spread through the population, perhaps eventually leading to a new species.

Thus, the fundamental issue at stake is the question of how *new* genes are created. The whole notion of sexual reshuffling of gene "snippets" is really secondary. It is in the *new* gene snippets that the fundamental changes in the blueprint of the individual are made. It is in the *new* genes, for instance, that the 54,000 new proteins are made that distinguish a human being from a fruit fly. How has the coding for making these new proteins been achieved in the evolutionary steps leading up to the human species?

The macroevolutionists do not have a mathematically formulated answer; they basically refer to the millions of years available for genetic innovation to take place. In essence, they wave their hands and accept their Darwinian paradigm of chance and natural selection as is. Theirs is an article of faith, no more, no less.

The microevolutionists, by their theory of mutations (discrete changes in the gene structure), open the door to a mathematical treatment of the problem of chance mutations. The chance argument legitimizes the use of probability theory in mathematically tackling the problem of creating new genes. However, this approach encounters numerous difficulties, as we shall see.

In 1986 well-known biochemist Robert Shapiro, professor of chemistry at New York University, published *Origins: A Skeptic's Guide to the Creation of Life on Earth*. In this book he calculates the number of chance trials that can have taken place in the course of life's evolution on Earth.

Presently, there exists on Earth a simple bacterium called *E. coli* that replicates once every 20 minutes. Shapiro considers a hypothetical precursor bacterium that replicates twenty times faster, or once every minute. A time span of 1 billion years gives evolution 5×10^{14} minutes, or time for 5×10^{14} trials on the structure of that single precursor bacterium. Shapiro assumes that these trials take place in an ocean that is 10 kilometers deep and covers the entire Earth. He then considers the ocean divided up into small trial volumes of 0.1 cubic millimeter, slightly larger than the size of a single bacterium. This implies that the ocean contains 5×10^{36} individual small volumes in which the trials take place and that each little volume can experience 5×10^{14} trials in the course of the allotted 1 billion years. The multiplication of these two large numbers, 5×10^{36} and 5×10^{14}, leads to the number of trials available for evolution to work on: Shapiro's number, or 2.5×10^{51}.

Shapiro's number is a purposeful exaggeration of the number of trials actually carried out to create this hypothetical bacterium. It is highly unlikely that such trials took place in every little microvolume each and every minute over the span of 1 billion years.[1] But for the first time, we have a realistic upper limit to the number of trials physically available on earth for the creation of life. Shapiro's number is so important because it demonstrates that the number of chance trials available is not infinitely large but instead is confined to a specific value. Mathematical calculations on the probability of success by chance trials have to be compared with this limited number of trials.

For the moment let us ignore all the more sophisticated gene creations needed to manufacture the more than 200 different types of body cells in the

---

[1] In this regard it is highly significant that realistically inferred mutation rates are far less than one trial per minute. In *Molecular Biology of the Cell* (p. 221) we learn that most mutations are damaging to the gene and that their negative effects have to be eliminated by natural selection. The estimated success rate is one constructive mutation per average gene in 200,000 years. Furthermore, the authors maintain that a tenfold higher mutation rate would limit the survivable genes in sperm and egg cells to the production of only 6,000 different proteins. "In this case evolution would probably have stopped at an organism no more complex than a fruit fly" (p. 221). The authors further argue that a tenfold higher mutation rate in the *body* cells would cause also a disastrous increase of cancer in these cells, thus destroying the individual. Apparently we have here a most significant limitation on how often changes in the genetic blueprint can be made at the molecular level.

human being, or the high-tech information put into the glorious structure of the brain, or the optical artistry of the eye. Let us consider just the simple bacterium *E. coli*. In order for the bacterium to function and replicate, science tells us that it needs an array of over 2,000 different enzymes. (We may recall that an enzyme is a convoluted protein that acts as a meeting place for the molecular participants in chemical reactions, thus speeding up the reactions by factors of 1,000 or more.)

A typical enzyme is generally made up of several hundred amino acids. In their book *Evolution from Space* (1981), Sir Fred Hoyle and Chandra Wickramasinghe conservatively calculate that the number of chance trials required to make an average enzyme is 10^{20}. This number is of course smaller than Shapiro's number, 2.5×10^{51}, but over 2,000 different types of enzymes have to be created within this total number of trials available.

Suppose that five different enzymes have to work together for the self-replication process, that is, the creation of a new individual bacterium out of the old one. In this case Hoyle and Wickramasinghe, using the established laws of mathematical probability, find that the expected number of trials necessary for success is $(10^{20})^5$, or 10^{100}, far more trials than Shapiro's number allows. Shapiro goes through a similar exercise to reach the same conclusion in his book (p. 298).

The Darwinian principle of chance creating genuinely new gene structures fails miserably. Many are the critics who have pointed out the flaws of a chance argument for the origin of life as we know it, going all the way back to Cicero in Roman times. But their voices have mostly fallen on deaf ears both inside and outside the Darwinian establishment.

As a result of the explosive advances in microbiology concerning the genome structures of various life forms and how information is coded into these genomes, a more realistic attitude is emerging that all is not right with the Darwinian chance argument.

The intellectual "life buoy" thrown to the classical neo-Darwinists is the emerging discipline of self-organization. That a life buoy is needed to keep classical Darwinism afloat is evident. Take, for example, quotes on the back cover of Stuart Kauffman's 1993 volume *The Origins of Order: Self-Organization and Selection in Evolution*. Various highly renowned macro-evolutionists therein acknowledge some of the weaknesses in current thinking about traditional neo-Darwinian views. Stephen Jay Gould of Harvard University, for instance, acknowledges bluntly that "Darwinian theory must be expanded to recognize other sources of order based on the internal genetic and developmental constraints of organisms and on the structural limits and possibilities of general physical laws," and he characterizes the profession as groping toward "a more comprehensive and satisfying theory of evolution." Similarly, John Maynard Smith of the University of Sussex voices a question now familiar to us: "Has there been time, since the origin of life on earth, for natural selection to produce the astonishing complexity of living organisms?"

What is it in the theory of self-organization that gives some neo-Darwinists hope for a way out of the current impasse in Darwinism as they see it?

For a better understanding of the concept of self-organization, we shall have to make a brief digression into physics. It has long been known to physicists—as in our everyday experiences—that matter (atoms and molecules of a given kind) can organize itself into different structural states depending on physical constraints and circumstances. A classic example concerns the three possible structural states of a substance like water. At temperature conditions above the boiling point, water exists in gaseous form as water vapor—a state in which the water molecules move relatively freely about. When the temperature is lowered below the boiling point, the water molecules in the vapor begin to organize themselves spontaneously into a more closely knit structure—that of a liquid assembly. The vapor condenses into its liquid water form as we know it in our daily experiences. The physicists say that the *collective emergent property* of the water molecules at this proper temperature is liquid. Likewise, when the temperature drops below the freezing point, the water molecules in the water liquid begin to self-organize into ice crystals; the collective emergent property of water molecules below 0 degrees centigrade is that of ice.

Over the past several decades, physicists have been able theoretically to explain to a certain degree why the water molecules are able to self-organize into these different structural states, given certain temperature and pressure conditions.

Some thirty years ago, certain microbiologists (for example, Nobel laureate Manfred Eigen in Germany and mathematical biologist Stuart Kauffman in the United States) began to speculate that the concept of self-organization in dead matter might be carried over into the realm of the biological sciences. Thus, the ancient Greek beliefs in the spontaneous generation of life were resuscitated, but at a vastly more complex level of theoretical analysis. Could it be that, given certain physical conditions of the environment, organic molecules like DNA, nucleic acids, proteins, and enzymes self-organized into life forms? This was the daunting question the theoretical biologists involved in the theories of self-organization asked themselves.

By far the most articulate exponent of the current theory of self-organization is Stuart Kauffman himself, currently at the Santa Fe Institute. In the introductory chapter to *The Origins of Order,* he reviews the conceptual outline of current evolutionary theory. Many of the weaknesses that we have pointed to in our critical review of neo-Darwinism are echoed there. In particular, Kauffman honestly faces Shapiro's dilemma in response to a famous assertion made by George Wald in the August 1954 issue of *Scientific American.* Wald's statement was as follows:

> One has only to contemplate the magnitude of this task to concede that spontaneous generation of a living organism is impossible. Yet here we are—as a result, I believe, of spontaneous generation. . . . Time is in fact the hero of the plot. The time with which we have to deal is of the order of 2 billion years. What we regard as impossible on the basis of human experience is meaningless here. Given so much time, the impossible becomes possible, the possible

probable, and the probable virtually certain. One has only to wait: time itself performs miracles.[2]

Kauffman then runs through the arguments leading to Shapiro's number and contrasts this number with Hoyle and Wickramasinghe's probability calculations. Although he admits to the failure of the Darwinian chance argument, he pins his scientific hopes on the theory of self-organization, especially in the particular direction his own research during more than twenty-five years has carried him.

His aim clearly "is not so much to challenge as to broaden the neo-Darwinian tradition" (p. 26). He aims to do this through studying the process of self-organization of the atoms and molecules that enter into the formation of living organisms. Rather than trying to break with the neo-Darwinists, he tries to meet traditional criticisms based on classical probability theory by showing the emergent properties of systems with large interacting numbers. Kauffman writes:

> I suggest throughout this book that many properties of organisms may be probably *emergent* collective properties of their constituents. The evolutionary origins of such properties then, find their explanation in principles of self-organization rather than sufficiency of time. (P. 22)

Kauffman's book basically enunciates an aspirational program. Kauffman hopes eventually to discover collective properties of atoms and molecules that may lead to an enzyme, a beak, an eye, a brain. In simpler language, it

---

[2]Even though this was Wald's opinion in 1954 and might be considered outdated, we find an echo of this article of faith still expressed in Stephen Jay Gould's *The Flamingo's Smile* (1985). In discussing Alfred Russel Wallace's philosophy, Gould writes: "He understood only too well that ordered and complex outcomes can arise from accumulated improbabilities" (p. 400). It is only fair to state that George Wald, in his later years, has come to embrace an altogether different concept that "this Universe is life breeding because the pervasive presence of mind has guided it to be so. . . . Mind and matter are the complementary aspects of all the reality" (1994, pp. 129–130).

would appear that Kauffman hopes to find that in the mathematical treat-ment of millions of interactions—in the collisions between certain atoms and organic molecules that are constituents of groups known as ensembles—an ordered pattern will crystallize that is stable in time and outside the realm of ordinary probability calculations.

A central theme in Kauffman's book is the study of the properties of spe-cial mathematical systems called Boolean networks. Among these, Kauffman selects for purposes of illustration one network having 100,000 components, for which each component receives two inputs from its surroundings. He is then able to show mathematically that this system "crystallizes" into 317 sta-ble, ordered states instead of the $2^{100,000}$ possible alternative states. Once this result is established, he urges, "Whatever else you mark to notice and remember in this book, note and remember that *our intuitions about the re-quirements for order in very complex systems have been wrong*" (p. 235, his italics). Exciting as this undoubtedly is as a finding on the behavior of a hy-pothetical mathematical system, it is hard to be convinced of its applicabil-ity to a system of 100,000 *organic* molecules in a cell or even in the primeval organic soup.[3]

Through the mathematical theory of self-organization, Kauffman hopes to find biological laws ordaining that an eye, for example, has to be formed rapidly in the course of evolution. He states, "Biology since Darwin is un-thinkable without selection, but may yet have universal laws" (p. 25).

This is a stunningly novel way to look at evolution, but how far has the theory of self-organization advanced in this ambitious program? After the details of Kauffman's book are scrutinized, a certain disappointment sets in. No self-organizing "recipes" or principles emerge that explain how a beak,

---

[3]Laboratory work on synthetic self-replicating molecules has recently been carried out by Julius Rebeck, Jr., and his colleagues at the Massachusetts Institute of Technology. (See July 1994 issue, pp. 48–55, of *Scientific American*.) However, the chemi-cal nature of these molecules is still different from those Nature employs in DNA (see Chapter 2). Furthermore, there still re-mains the fundamental problem of inserting a massive information content into these synthetic self-replicating molecules in the course of evolution.

an eye, or even a single cell inevitably arises out of an ensemble of organic molecules.

Although Kauffman's book conveys in an exciting manner the present status of the theory of self-organization, the inescapable conclusion is that there is still a long way to go before we can take seriously the applications of that theory to biological evolution. Even Kauffman himself admits in his epilogue that "it is no rhetorical apology to recognize the inadequacy of the efforts undertaken here. The themes are large and new and remain incompletely articulated" (p. 644).

Thus, Kauffman's expanded neo-Darwinism still leaves us with an article of faith—albeit not a vaporous faith in time as a biological "miracle worker," but rather faith that the ongoing mathematical applications of the theory of self-organization will eventually lead to the discovery of biological laws for collectively ordered states.

An apparent difficulty with this approach is that the world of ideas must be part of these ordered states and evolving idea structures (for example, the eye). How will Kauffman move from treating a Boolean system of 100,000 abstract units to treating a real-world sequence of 100,000 genetic code letters in the organic DNA molecules of a rosebush—a sequence in which is coded the information to create a rose flower? The evolving natures of idea structures imply that Kauffman's biological "laws" must themselves be evolving, unlike the stable, unvarying physical laws.

The incorporation of ideas into the organized structures of life forms implies the insertion of *information* into their DNA gene structures. Even a macroevolutionist like Mayr acknowledges the emphasis on information accumulated in the genetic constitution of an organism. In *Toward a New Philosophy of Biology* we read:

> Where organisms differ from inanimate matter is in the organization of their systems and especially in the possession of coded information. . . . There is nothing in any nonliving (except man-made) system that corresponds to the genotype, a system that has selectively stored vital information during the billions of years that life has existed on earth. (P. 2)

The total amount of information stored in the genotypes of all species present and past is staggering. Biologists currently estimate that there are more than 3 million different species on Earth. Conservatively estimated, there may well have existed 2 billion species in the history of life on Earth (R. C. Lewontin in *The Fossil Record and Evolution*, p. 17). If we grant only the equivalent of two typewritten pages of new information for the creation of each new species, the total information content in all life forms past and present comes to about 4 billion pages, or 10 million books of 400 pages each. Alternatively, if we posit that each species past and present is characterized by the creation of only one new gene, the evolution of life on Earth implies the creation of 2 billion new genes in the course of 600 million years.

We must keep in mind that the total number of trials available on our planet is at most a few times Shapiro's number. Earlier we saw that this number of trials was insufficient even to make only a few connected enzyme structures. How, then, to account for the creation of 2 billion new *genes*

Lately, there have appeared interesting developments in several new scientific disciplines—information theory, chaos theory, and the theory of self-organization—that have applications to the problems of evolutionary biology. Some representative samples of the literature, besides Kauffman's book, include Jeffrey S. Wicken's *Evolution, Thermodynamics and Information*; Ilya Prigogine and Isabelle Stengers's *Order out of Chaos*; Manfred Eigen's *Steps Towards Life*; and Herman Haken's *Information and Self-Organization*. However, although some of the results in those fields are impressive, we are left with the distinct feeling that those disciplines have a long way to go before they can explain the creation of even a single cell with its attendant thousands of chemical reactions.

The present-day intellectual situation in evolutionary biology bears certain resemblances to earlier paradigm developments like, for example, those in astronomy. The pre-DNA days of purely observational evolutionary biology can be viewed as equivalent to the Ptolemaic epoch, whereas the advent of molecular biology has ushered in a Keplerian phase of mathematization on an information/theoretical basis, demanding an explanatory paradigm of a Newtonian type, which transcends the current Darwinian concept of chance mutations.

In *The Structure of Scientific Revolutions,* Thomas Kuhn scrutinizes paradigm changes in the history of modern Western science from the time of Galileo. In a chapter entitled "Anomaly and the Emergence of Scientific Discoveries," Kuhn maintains that discovery, as he sees it,

> . . . commences with the awareness of anomaly, i.e. with the recognition that nature has somehow violated the paradigm-induced expectations that govern normal science. It then continues with a more or less extended exploration of the area of the anomaly. And it closes only when the paradigm theory has been adjusted so that the anomalous has become the expected. Assimilating a new sort of fact demands more than an additive adjustment of the theory, and until that adjustment is completed—*until the scientist has learned to see nature in a different way*—the new fact is not a scientific fact at all. (Pp. 52–53, italics added)

The Darwinian anomaly is represented by the vast information content deposited in the total sum of life species past and present and the inadequacy of the Darwinian chance argument to account for this mathematically.

However, when a new paradigm emerges, not all aspects of the old paradigm will necessarily be eradicated. In any fundamental change of Darwinian theory, certain original idea structures will of course be retained.

1. The overarching concept of an evolution of life forms, from the primitive single cell to the variegated multicellular life forms, is indisputably here to stay.

2. The concepts of population dynamics and population genetics, whereby the individual species through sexual recombination has an evolving gene pool in terms of secondary characteristics (such as beak size or colors), will also be retained.

3. The role of natural selection (therein including the effects of the environment) in "fine-tuning" the species' adaptation will be retained.

And, of course, within this context, the natural physical laws have to be obeyed. However, at the roots of this hierarchical structure lies the problem of speciation, with the coding of new information into genes. The central issue is whether chance—or, for that matter, self-organization—is an adequate mechanism for inscribing into the genetic DNA structures the extraordinarily complex information displayed in the structures of living organisms.

We may find that Alfred Russel Wallace, one of Darwin's contemporaries, will come to outshine Darwin as our view of evolution expands. In *Man's Place in the Universe: A Study of the Results of Scientific Research in Relation to the Unity or Plurality of Worlds* (1903), published when Wallace was eighty, Wallace makes two summarizing statements on the creation of life and its evolution. The first statement concerns the Darwinian conception of chance: "One considerable body, including probably the majority of men of science, will admit that the evidence does apparently lead to this conclusion, but will explain it as due to a fortunate coincidence." Wallace then goes on to contrast this view with his own personal conviction: "The other body, and probably much the larger would be represented by those who, holding that mind is essentially superior to matter and distinct from it, cannot believe that life, consciousness, mind are products of matter. They hold that the marvelous complexity of forces, which appear to control matter, if not actually to constitute it, are and must be mind-products." (Quotes found in Gould, *The Flamingo's Smile*, pp. 400, 401.)

We should consider the words of renowned natural philosopher Francis Bacon (1561–1626), who in *Novum Organum* wrote, "For the world is not to be narrowed till it will go into the understanding (which has been done hitherto), but the understanding is to be expanded and opened till it can take in the image of the world." It is with that conviction that we shall enter the world of the humanists and listen to some of their voices, which echo up through the centuries and millennia and speak of a *Designer*.

Just as the idea of the atom lay dormant for more than 2,000 years before it took hold in Western scientific minds, so there may exist other dormant "thought seeds" that might help us out of the present impasse in evolutionary biology. Such a thought seed is the idea of pantheism: God and Nature intertwined.

"The Gods will not speak to

us face to face until

we ourselves have a face."

—C.S. LEWIS (1898–1963),
English novelist and essayist

The IMPLICATIONS
of BIOLOGICAL DESIGN

As we have seen in earlier chapters, the human species, throughout its existence, has imitated biological designs that it observed in Nature. With the leisure time afforded by the rise of the tribal cultures and the later major civilizations, human beings also began to speculate on the *origins* of such biological designs or forms.

In this chapter we shall attempt to give a brief outline of the development of pantheistic philosophy in Western civilization since the golden age of Greek culture. We will discover that the Greeks' idea of God and Nature intertwined survives on through the millennia, periodically being reevaluated and rephrased in terms of the accumulated knowledge of the period in question.

Ours is an age in which such a reevaluation seems particularly appropriate in view of the spectacular advances in microbiology and theoretical physics and the resultant problems for the Darwinian paradigm.

The seeds of Western natural science are found in Greek civilization, as it developed from the sixth century B.C. until the third century B.C. And curiously enough, it is there that we find the first coherent philosophical speculations about biological design and the evolution of life forms in the writings of Aristotle (384–322 B.C.) and of Democritus (ca. 500–404 B.C.), the father of the concept of the atom. From those two great Grecian minds originated the ever-present confrontation between vitalism versus mechanism.

In the excellent overview *Mechanism and Vitalism* (1962) by Ranier Schubert-Soldern, an Austrian professor of theoretical biology, we find the following distinction between the vitalist and the mechanist:

> The vitalist sees in the living organism the convergence of two essentially different factors. For him matter is shaped and dominated by a life principle; unaided matter could never give rise to life. The mechanist on the other hand, denies any joint action of two essentially different factors. He holds that matter is capable of giving rise to life by its own intrinsic forces. The mechanist considers matter to be "alive." The vitalist considers that something immaterial lives in and through matter. (P. 10)

The concepts of both vitalism (entelechy) and pantheism (God in Nature) were joined in Aristotle's philosophy of life forms. Aristotle essentially employs a macroscopic (nonatomistic) point of view, and the concept of *form* is an essential cornerstone in his philosophy. All bodily objects in the Universe are, to his thinking, made up of matter and form. Form is imprinted upon matter in a variety of ways, as much in "dead" forms as in "live" forms. Water, for example, can appear in the form of ice or liquid or gas. Water, in Aristotle's terms, would be *materia prima* (primary quality of matter). As soon as it receives an imprint (for example, by becoming ice), the water appears as *materia secunda* (secondary quality of matter). Those are elements of Aristotle's theory of matter and form (hylomorphism). Form, as such, is called *forma substantialis* (object form). Carried over into the realm of the human artisan, say a potter working in clay, clay would be the *materia prima* out of which the potter would express *forma substantialis* (in this case an artistic idea).

Forms can be dissolved and transformed into new forms; thus, a new form can be imprinted on matter already having a form. The cow eating grass changes the form of grass into substances from which she derives energy, builds new tissues, or creates a new calf.

Aristotle holds this hylomorphic concept applicable to all living and all nonliving bodies. However, he believes the form imprint on living bodies is caused by an active principle that turns what is possible into what is real. This active principle, entelechy, according to which *materia prima* tries to actualize a form imprinted upon itself so as to yield life forms, relates to Aristotle's concept of *teleology* (the doctrine that phenomena are not only guided by mechanical forces but that they also move toward certain goals of self-realization). To Aristotle, the idea of form is latent within the life form (as the adult is latent in the embryo), not outside it. Thus, evolution proceeds not because of material forces pushing on matter from the outside but because of latent ends that lie embedded in the matter of the life form. Aristotle considers evolutionary change as a motion toward an ideal. With Aristotle originated the idea of a grand design caused by a divine intelligence, his concept of a God.

In *De Anima* Aristotle states that "the soul is the first grade of actuality of a natural body having life potentially in it." In his later years, he became convinced that a part of the soul comes from outside of the life form and is a part of pure reason. When a man dies, that part of his soul will go back to the universal reason (God) that absorbs it.

Aristotle is rightly called "the father of Western empirical science" because he firmly believed in a rational, ordered approach to unveiling truths by observation and experimentation. Yet there were decidedly mystical elements in his approach to natural science. His concepts of entelechy and teleology may yet have their impact on future developments in the thinking behind evolutionary biology.

Democritus came to embody the mechanistic view of life forms, as he elaborated the atomistic concept—first enunciated by Leucippos (around 500 B.C.)—wherein all natural processes are interactions between the smallest constituents of matter, the atoms. According to Democritus, matter is built up of various kinds of atoms, and the variety of material things is produced by different combinations of those atoms. Yet, when it came to producing life forms, Democritus felt he had to resort to the postulation of special "soul" atoms that animate ordinary atoms into life forms. The soul atoms were lighter and subtler than the ordinary matter atoms and were immersed in all material forms. To Democritus, the whole Universe was animated (psyche-filled), but there was no God. Everything was mechanistic and deterministic.

Finally, Democritus conceived of the eidola, the very lightest group of atoms, which were scattered everywhere and were capable of influencing the fate of humanity. Such atoms were apparent in foreboding dreams, visions, and divinations. Thus, although it was all mechanistically ordered, Democritus's Universe was by no means devoid of idealism.

The intellectual legacy of Aristotle and Democritus was carried into a culturally awakening Western Europe in the age of scholasticism. The great scholar Thomas of Aquinas (1225–1274) incorporated the hylomorphism and teleology of Aristotle into the Christian worldview, where they were safely lodged until the scientific intellectual breakthroughs of the nineteenth century. The atomistic concepts of Democritus were the first to come to full bloom in Western science with the appearance in 1808 of John Dalton's pathbreaking treatise in chemistry, *A New System of Chemical Philosophy*. Although scientists in the seventeenth century, like Pierre Gassendi, Robert Hooke, and Isaac Newton, had been acquainted with the atomistic theory, it fully entered the mainstream of Western scientific thought with the advances in chemistry spawned by the Greek atomistic ideas. However, Democritus's concepts of soul atoms and the subtle eidola atoms were conveniently forgotten by the mechanists, since there was no apparent use for those ideas.

In light of ultramodern molecular biology, it is fascinating to see how close Aristotle came in his concept of entelechy to enunciating the notion of a genetic code as we conceive it today. The genetic code in a cell's DNA is nothing but a prescription, an actual blueprint, for translating potential form into actual form. The activating principle in starting the genetic code toward building the actual form, however, remains to be worked out.

Thus, we may be able to discern in molecular biology today that two mainstreams of Greek thought seem to merge. On the one hand, Democritus's atoms are the building blocks for the genetic code. And on the other hand, there is part of Aristotle's entelechy, which forms the genetic blueprint that uses a sequential arrangement of Democritus's atoms by way of the nucleic acid letters in the genetic code.

In evolutionary biology today, as it may have appeared to Aristotle, there is an evolution of simple life forms into more complex ones. Aristotle was a great compiler of scientific information in all fields, including geology and botany, and could not have escaped noting the enormous differences in the complexity of existing life forms, with humans at the top. His enunciation of the principle of teleology (progression toward an ideal form) fit the empirical observations at the time. That progression involved the concept of a higher, godlike intelligence guiding and shaping the ascendancy toward an ideal form. Thus, through Aristotle arose another mainstream of thought in antiquity: pantheism, God as revealed in Nature.

The word *pantheist* was not introduced until 1705 by British philosopher John Toland in his work "Socinianism Truly Stated," but the concept of a divine power that infused some of the material objects with life and movement arose early in Greek philosophical thought, albeit articulated in a somewhat indistinct manner.[1] Thinkers like Xenophanes (around the sixth century B.C.) and much later Marcus Aurelius (A.D. 121–180) reverted to a form of pantheism as a reaction to the "vulgar" polytheism espoused by Homer, Hesiod, and others with a panoply of anthropomorphic gods involved in humanlike activities. But not one of them made a clear distinction between the assertion that an object is "divine" or that an object is "informed" or "animated" by a divine power.

---

[1] At this point it is important to emphasize that in this book we are mainly concerned with the idea of pantheism and how, in light of our modern knowledge of physics and biology, it can be applied to evolutionary biology. We will not consider the concept of vitalism any further, nor that of panpsychism, because these are essentially a subset of—and thus a part of—the idea of pantheism.

In the golden age of Greek philosophy, Plato (428–348 B.C.) emerged—preceding Aristotle by a generation—and laid the basis for an everlasting philosophical empire that has existed as a counterpart to Aristotle until today. To Plato, we owe the concept of another reality—the world of ideas, perfected ideals for which the material forms are imperfect renditions. Whereas Aristotle, the pragmatic empiricist, focused on rational studies of the forms, Plato was an abstract theoretician who found the greatest pleasure in the pure theoretical contemplation of ideas. Experimentation and observation had little value in Plato's world of philosophy.

Plato's view of the creation of the world set forth in his dialogue *Timaios* is not pantheistic but is a derivative of the Pythagorean natural philosophy. In *Timaios,* the Demiurge creates once and for all the various life forms. There is no concept of evolution as in Aristotle's later teachings.

What is interesting in terms of modern evolutionary biology is that Plato's theory of a world of ideas points to a kind of repository of information content that is drawn upon from outside the created life forms. In our previous chapters, we have consistently pointed to the sophisticated knowledge that has gone into the creation of life forms. That information content does not appear to reside in Democritus's concept of the atom.

Plotinus, who lived at the beginning of the decline of the Roman Empire in the third century A.D., is the founder of the neo-Platonistic school of philosophy and occupies a singular transitional position by assimilating, deepening, and elaborating on concepts found in the philosophies of Plato and Aristotle.[2] He amalgamated those concepts into a rather consistent philosophical system, which is eminently pantheistic. But he also significantly introduced a mystical element—ecstasy, or stepping outside of oneself—that attracted the Christian theologians. Plotinus's philosophy, when taken over by such Christian thinkers as Saint Augustine and later Thomas Aquinas,

---

[2]This exposition of Plotinus's philosophy owes much to a review article by Philip Merlan in *The Encyclopedia of Philosophy.*

ensured that the Platonic and Aristotelian legacies of Greek culture were incorporated and continued in the mainstream of Western Renaissance thought.

Plotinus's teachings have had an immense value in Western civilization. As Bertrand Russell, in his *History of Western Philosophy*, puts it:

> Like Spinoza he has a certain kind of moral purity and loftiness which is very impressive. He is always sincere, never shrill or censorious, invariably concerned to tell the reader, as simply as he can, what he believes is important. . . . Whatever one may think of him as a theoretical philosopher, it is impossible not to love him as a man. (P. 310)

Plotinus was a sage and a saint in his attitude toward the worldly goods of life and in his comportment toward his fellow man. The pantheistic ideas inherent in his thinking issued from one of the major philosophers in Western intellectual history, Aristotle, and as such he deserves to be taken very seriously in our quest for thought seeds we might apply to the realm of evolutionary biology.

Plotinus's philosophy operates with a trinity of fundamental concepts: the One, Intelligence (Nous), and the Soul. But unlike the Christian trinity, those concepts are hierarchically ordered.

The concept of the One, which issues from Plato in unelaborated form, is considerably elaborated by Plotinus. He sees it as the highest principle the cause of everything that is, but as such it is above being and quality. It may also be considered the object of universal desire, the Good. Because it is the One, it is without multiplicity and awareness. There is nothing of which the One is cognizant, and there is nothing of which it is ignorant. Those apparently self-contradictory statements in Plotinus's philosophy on the nature of the One present a difficulty of description shared by many religions and philosophies. As Bertrand Russell puts it: "The One is indefinable, and in regard to it, there is more truth in silence than in any words whatever" (1946, p. 288).

The realm of the One is followed by the realm of Intelligence, Plato's world of ideas, which Plato took as the realm of true being. Plotinus instead conceives of ideas as the thoughts of the One. Intelligence emanates from the One and is frozen in its flow by a contemplation of the One by Intelligence.

Light is a very important ingredient in the philosophies of both Plato and Plotinus, and frequent references are made to the Sun, which is both a light-giver and the "What is lit." Allegorically, Intelligence may be thought of as the light by which the One sees itself. Despite its multiplicity (of ideas), Intelligence retains its unity, much as, say, the discipline of mathematics is a unity but contains a multiplicity of individual mathematical theorems.

From Intelligence emanates the third and lowest realm, that of the Soul. That realm, like Intelligence, involves a multiplicity of smaller individual entities—not ideas, as in the realm of Intelligence, but souls. Some souls do not enter into material bodies but stay apart from matter. Others enter into celestial bodies like the Sun, Moon, planets, and stars. Still others enter into terrestrial bodies and life forms. Being incarnated into celestial bodies offers no resistance to the souls, so the souls do not feel imprisoned. But those souls in the terrestrial bodies do feel trapped, and they may become alienated from Intelligence.

Just as the flowing emanation of Intelligence is frozen by the Intelligence contemplating its One source, so the emanation of the Soul from the realm of Intelligence is arrested by the Soul's contemplation of the realm of Intelligence. However, in the Soul, multiplicity dominates over unity, so the realm of the Soul is of lesser perfection than that of Intelligence.

From the Soul emanates matter, which in Plotinus's philosophy is totally indetermined (undefined) and remains totally unaffected by the incorporation of ideas. Matter illuminated by the Soul becomes the physical world, the model of which is in the realm of Intelligence. Here Plotinus elaborates and deepens Plato's allegorical divine artisan, the Demiurge. Thus, Plotinus says, the Soul mediates the imprinting of ideas onto the matrix of indeterminate matter. In modern terminology, we could say that the Soul writes the genetic code messages. The Soul of Plotinus forms a part of Aristotle's entelechy, the active principle, imprinting form upon matter.

Plotinus's concept of cosmic sympathy relates to his concept of the world as one animated organism.[3] The parts of the world communicate with each other exclusively through mutual sympathy. Plotinus draws an analogy to the sympathy between parts of an organism, which live in harmony and coordination but whose mutual "affection" need not be perceived by memory and sensation.

With his strong sense of the sensory beauty of the physical world, Plotinus combats the Gnostics' pessimistic world attitude. The Gnostics considered the world to be a hostile place created by an evil god. In their view, humans harbor a spark of the original divine substance, but this spark is darkened by the material in which it is imprisoned. Plotinus refuses to concede a "debasement" of the individual soul in its material body form. As his only innovation over Plato, he distinctly claims that the soul, even when incarnated in the body, leads its unpolluted "celestial" life unseparated from Intelligence. That claim implies an assumption of an unconscious part in us, where the "hidden" life of the soul resides.

In his later life, Plotinus embraced the concept that man's true self is represented by the presence of Intelligence in him and that man's true happiness lies in the exercise of his true self by the act of contemplation. That act, in exceptional cases, can lead to a knowledge of the One—a mystical state that Plotinus repeatedly attained, according to the writing of his student Porphyrios.

In the fifth volume of his *Enneads*, Plotinus sets forth one of the most explicit and impressive descriptions of the upward journey of the Soul to reach ecstatic union with the One, described by the statement "through light light." In Plotinus's final words:

---

[3]We note here in embryonic form the idea of the interdependence of all life forms on this planet, as argued 1,700 years later by Lovelock and Margulis in their Gaia hypothesis.

But how is this to be accomplished?

Cut away everything.

We thus experience in the personality of Plotinus an extraordinary union of a profound rational thinker and a genuine mystic. The powerful attraction of his teachings affected a long line of Christian philosophers from Saint Augustine to Thomas of Aquinas.

Since Plotinus's philosophical lifework, no comparable intellectual achievements in the domain of pantheistic philosophy emerged until the appearance of Benedict de Spinoza (1632–1677) in the seventeenth century. He was a contemporary of Isaac Newton (1642–1727) and Gottfried Leibniz (1646–1716). Bertrand Russell singles out Spinoza as, aside from Plotinus, "the noblest and most loveable of all the great philosophers" (1946, p. 592).

Like Plotinus, Benedict de Spinoza led a quiet, contemplative life. Spinoza never founded any school or had any famous students.[4] He was offered a professorship in philosophy at the University of Heidelberg in 1637 but declined it because he thought it would deprive him of independence and tranquility. Instead, he chose to support himself by grinding lenses and discussing philosophy with friends and in correspondence with others.

Spinoza's exposition of his philosophy stands in absolute contrast to that of Plotinus. Whereas Plotinus adopts the Grecian mode of discursive, even rambling discourses on his subject (not unlike Plato or Aristotle), Spinoza adopts an extremely rigorous, logical structure of propositions, particularly in his *Ethics*. Spinoza's propositions are patterned after Euclid's geometry, with its interlinked axiomatic logical propositions. And whereas Plotinus attains the supreme mystical union with the One, Spinoza's ultimate goal is the "intellectual love of God," which frees man from the vagaries of chance and

....

[4]This exposition of Spinoza's philosophy owes much to a review article by Alasdair MacIntyre in *The Encyclopedia of Philosophy*.

circumstance; all emotions are to be replaced by "adequate" ideas, ideas that are "true" and "valid" independent of the idiosyncrasies of individual men. Nevertheless, both Plotinus and Spinoza are pantheists, though Spinoza comes to this state through a hard-headed rational metaphysics.

In Spinoza's metaphysics, *God* and *Nature* are but two names for a single whole system. By a series of logical propositions, Spinoza deduces that there exists only one unique, single substance that underlies all reality. Because *God* is the name of the one substance, whose other name is *Nature,* the contrast between God and the world is obliterated. God *is* this world and more, and as a result God is the immanent and continuing cause of this world and not its transient first cause. This reasoning represents a far more complete and undiluted pantheism than that of Plotinus. And it stresses a time factor in a very important way, which opens up for God's participation in the evolution of life forms.

However, Spinoza adamantly rejects the proposition that God has any purposes, designs, or desires for the world. Spinoza is driven to this extreme position by his logic. The basic argument is that God, if He were to have purposes, designs, or desires, would strive for something that He lacks. And this to Spinoza is absurd, because by his definition of God, God can lack nothing.

It is thus into a starkly deterministic Universe that Spinoza's metaphysics leads us, with no place for a free will. *Everything* happens because of God's inscrutable will.

Spinoza defines *attributes* as "that which intellect perceives in substance as constituting its essence" (*Ethics* I, Definition IV), and God possesses an infinite number of attributes, of which mind and matter are two we are given to perceive. This implies, according to Spinoza, that mind (the world of ideas) and matter (the world of physical bodies) are but two different ways of envisaging the same underlying reality (God). Thus, reality can be viewed as a variety of physical bodies causally connected *or* as a series of ideas that may be logically connected. The two networks (physical bodies and ideas) will correspond exactly because they represent the same reality viewed in two different ways.

In pursuing that "mirror-image" structure of ideas versus bodies, Spinoza finally constructs a hierarchy of ideas, where at the bottom reside the physiologically produced images. Above them lies the domain of "adequate ideas," which comprise common notions shared by all men. Those adequate ideas form the elements of natural sciences, thus forming part of a logically interlocked system in which the necessity of the relationship of every item to every other item is manifest at the top of the hierarchy. The total system of ideas is the infinite idea of God, and only God possesses a totally adequate idea of himself. Insofar as man approaches the possession of such an idea, he necessarily approaches the condition of God and becomes God to some extent. That third and highest grade of knowledge is that of the divine mind, which Spinoza calls intuitive knowledge.

Spinoza's thought structure rises majestically like the cathedrals of the Gothic Age but built with stones of logic. It firmly wedges the principle of Nature as totally immersed in a principle of God, with nothing outside it. It provides a temporal quality to the coextension of God and Nature.

As we shall see in the next chapter, there may be distinct tangential points between some of Plotinus and Spinoza's pantheistic ideas and certain controversial branches of the Western philosophy of physics. Certain quintessential ideas of Plotinus and Spinoza appear ready to be incorporated into evolutionary biology, but they need to be rephrased in the context of our modern knowledge of physics, biology, and even psychology. In the last chapter of this book, we shall consider making such an update.

Just as the philosophies of Aristotle and Plato have periodically been reevaluated by succeeding generations within the context of intellectual advances, so have the monumental pantheistic thought structures of Plotinus and Spinoza. Although the two philosophers are distant to us in time, both have continued to influence modern thinkers, including famous French philosopher Henri Bergson and, later, eminent mathematician and philosopher Alfred North Whitehead.

Henri Bergson was born in Paris in 1859. As a high school student he was deeply interested in both classical and scientific studies. While attending the famous Ecole Normale Supérieure, he excelled in mathematics and wrote a distinguished essay on a problem in Pascal's geometry. In 1900 he became professor of philosophy at the prestigious Collège de France in Paris, was elected to the French Academy in 1918, and received a Nobel Prize in literature in 1927. In 1921 he left teaching and became involved in politics and international affairs with the intention of promoting peace and cooperation among nations. He died in Paris during World War II in 1941.

Early in his life, Bergson became inspired by the teachings of Plotinus in philosophy[5] and by the works of Darwin in evolutionary biology. We shall examine Bergson's philosophy chiefly in the context of evolutionary biology, although his writings encompass a wide variety of subject matter, from a critical analysis of the phenomenon of laughter to the phenomena of memory and time.[6]

In Bergson's thought structure, for the first time ever, the strong current of pantheism running through Western philosophical thinking becomes amalgamated with elements of Darwinian findings into a consistent theory of evolution under the influence of the scientific findings in biology, paleontology, and even psychology.

That revision of pantheism is of utmost significance for the future incorporation of pantheistic conceptual structures into the conceptual framework of the natural sciences. Spinoza believed that a self-consistent structure of pure logic could be erected to explain the phenomenon of life and existence and that this structure could be divorced from empirical findings. However, it appears today that the pantheistic concept can come to genuine flowering only if conceived of as an extension of existing paradigms in the natural sciences. We will vigorously pursue that idea in Chapter 11.

---

[5]See reference to a lecture series on Plotinus that he gave at Collège de France, 1897–1898 (*Creative Evolution*, p. 384).
[6]The following exposition of Bergson's philosophy owes much to a review article by T. A. Goudge in *The Encyclopedia of Philosophy*.

In the human species has evolved an ability for conceptual and rational thought, which is traditionally known as the intellect. This capacity, according to Bergson, arose in humans because they (1) are one of the social animal species, (2) are tool-forming animals, and (3) have the most complex language structures of all animals. Thus, the intellect arose out of practical biological needs.

This biological usefulness of the intellect also carries its own limitations and restrictions. Bergson forcefully attacks the belief that the exclusive application of the intellect in the natural sciences leads to a complete understanding of reality. Originally, the intellect tended to break up the external world into separate objects in space and time, so as to serve biological needs. But when this tendency is carried over into the natural sciences, it leads to an artificial breaking up of experiential reality into a movielike representation of "still images," which is then analyzed in terms of logical structures such as mathematics. Thus, fictitious structures of thought, such as atoms, are created to represent some aspects of reality.

According to Bergson, besides the intellect there exists another vital capacity in the human mind termed *intuition*. This capacity arose evolutionarily out of the instincts, as exemplified most vividly in social insects. According to Bergson, intuition is instinct "disinterested" and decoupled from the needs of action and social life. It is somewhat like an artist's ability to see the world in a purely sensory manner without the intervention of thought. However, intuition yields knowledge rather than aesthetic experience, a knowledge that later may be conceptualized by the intellect.

Intuition enters into what it knows; being based on direct experience, it dispenses with symbols (abstract ideas) and produces knowledge that is absolute. Intellect remains outside what it knows, requires symbols, and produces knowledge that is always relative to some viewpoint.

The intellect has also been used to construct elaborate systems of metaphysics, which were concerned with the realm of the spirit. Bergson proposes to reject these metaphysical systems of the intellect and to replace them with a new kind of metaphysics based on the faculty of intuition. This new intuitional metaphysics would reveal another aspect of reality as reflected in the realm of the spirit. And thus the two disciplines of metaphysics and natural sciences would give complementary perspectives of reality. The new metaphysics based on intuition would supplement the sciences, especially by giving a true account of duration (as personal experience o the passage of time) and of becoming (that is, evolving).

Darwin's *On the Origin of Species* was published the same year that Bergson was born, and Darwin's work permanently affected the direction of Bergson's philosophical thinking. Bergson accepted the reality of historical evolution, but he rejected various explanations for how it happened, such as those put forward by Darwin and Lamarck. He felt that there must be more to the phenomenon of evolution than a mere material rearrangement of atoms and molecules. In this view he was influenced by the European Western tradition of vitalism but also by the works of Plotinus.

Bergson's thoughts on evolutionary matters were laid down in an immensely popular book, *L'Évolution Créatrice* (Creative Evolution), published in 1907, two years after Einstein's special theory of relativity. The book was widely acclaimed and read by a number of influential thinkers in France, as well as Great Britain and the United States. Famous American philosopher and psychologist William James enthusiastically embraced Bergson's ideas and wrote to the author, "Oh, my Bergson, you are a magician and your book is a marvel, a real wonder."

However, the intellectual stir created at the time Bergson's text was published has since subsided. Today Bergson's name is rarely mentioned, and his works are often forgotten. In a deep sense Bergson's book has yet to be fully appreciated, and in the long run it may turn out to be of far greater importance than Einstein's theories.

Bergson points out that Darwin's theory fails to take into account a number of facts. In particular, it fails to account for how a complex organ like the vertebrate eye has evolved. We recall that in Darwin's explanatory scheme, there occur in every population of a species a number of random variations that have a maximum adaptive value, which ensures the survival of that particular individual and thus the evolution of its species. However, as we have discussed, the mammal eye is made up of a number of coordinated parts. If one of these parts happened to vary independently of the rest, the result could adversely affect the whole functioning of the eye.

It is a fact that the evolution of the eye has occurred, so each stage of evolution must have happened in a coadapted manner with respect to each part, allowing the organism as a whole to continue functioning. But this goes counter to Darwin's theory of random variations of the individual parts. Some unknown agency must have kept the organism functioning through its evolutionary development.

Another fact that perplexes Bergson is the evolutionary development toward greater complexity in life forms. The first prokaryotic cells were relatively simple structures but were extremely well adapted to their environment. Why did the evolutionary process not stop at this stage? Why did life evolve into the multicellular life forms, whose complexity implied greater vulnerability to malfunctioning and death? Bergson again considers the explanation of random variations and natural selection to be insufficient.

Instead, he brings to bear his thoughts on what intuition reveals of us as human beings. Because we are inextricably parts of life on Earth, the forces at work in us should also work in all life forms. Thus, our personal sense of duration and becoming transcends mere scientific explanations. Furthermore, we are imbued with a distinct consciousness of something driving our personal evolution in time. This "something" Bergson calls élan vital (vital impetus), an agency that permeates the whole evolutionary process and causes its dominant features.

Bergson's concept of the vital impetus received much praise and criticism. Bergson himself kept a certain sober distance from his concept. In fact, in his book on creative evolution he says that although "the vital principle may indeed not explain much, it is at least a sort of label affixed to our ignorance, so as to remind us of this occasionally, whereas mechanism invites us to ignore that ignorance" (1911, pp. 48–49).

As we shall also see in the case of the process philosopher Alfred North Whitehead, Bergson struggles to formulate new concepts that can enlarge our understanding of the evolutionary processes. Such a struggle by necessity entails a certain "fuzziness" and lack of clarity. The concept of élan vital must be considered a seed idea, incomplete in structure and form. We shall see in the last two chapters how this idea may be rearticulated in a new form—that of the Planetary Mind Field—and integrated into the natural sciences of *our* times.

Nevertheless, Bergson cannot refrain from speculating on how this élan vital operates and what it relates to in our human experience. Believing that spirit and matter represent two aspects of Reality (an attitude similar to that of Spinoza), Bergson naturally conceives of the élan vital as a "current of consciousness" that has penetrated matter. Given the right physical circumstances in the planetary environment, the élan vital gives rise to life forms and directs the course of evolution.

Bergson does not fall into the trap of speaking of a special vital energy nor of a preconceived plan for evolution. The élan vital does not add any special energy to that already residing in matter itself. The vital impetus is not finalistic, but it does build upon earlier achievements, thus engendering progress.

So, in Bergson's thinking, the appearance of the human species is *not* preplanned in evolution, but neither is it wholly accidental as it grows out of earlier evolutionary developments. It is an ultimate manifestation of the creative aspirations of the élan vital. In humanity the level of consciousness is raised to its highest point, resulting in a maximal dominance over matter.

According to Bergson, matter and spirit coexist and are interdependent everywhere in the Universe. Thus, the vital impetus is not limited to the Earth; creative evolution is a cosmic process that may well result in sentient beings elsewhere but not necessarily in human form.

In his text on creative evolution, Bergson speaks of the vital impetus as a "supraconsciousness" (*Creative Evolution,* p. 284). He even attaches the name of God to the vital impetus (p. 271): "God thus defined, has nothing of the already made; He is unceasing life, action, freedom." However this God concept breaks with the fundamental beliefs in traditional Western theology. Bergson envisions a God that is neither omnipotent nor omniscient but rather is pure spiritual activity struggling to manifest itself in a material world and thus is limited.

In Bergson's view, evolution continues now to complete the development of our human species. That future development will depend upon the evolution of human societies, because humanity is in part a social animal. In one of his last works, *The Two Sources of Morality and Religion* (1935), Bergson distinguishes between two kinds of societies: closed and open.

According to Bergson, closed societies are to a great extent shaped by the intellect. They are self-centered, conservative, and authoritarian. In order to preserve this stability, they often resort to war. They embrace a static, nonevolving morality and a closed religion that is dogmatic and ritualistic. The individual within these closed societies is forced to conform by the pressure of the community.

Bergson asserted that such closed societies are obstacles to human evolution. After his early retirement from the prestigious Collège de France in 1921, he devoted himself to a number of political and international engagements, heading diplomatic missions to Spain and the United States. Largely because of these experiences, he believed that the next societal development would be the establishment of an open society that would embrace all of humanity. Such an open society would entail open and flexible morals and religions, enabling a maximum of freedom of expression among individuals. As part of the evolutionary development, a new religion would emerge, free of the dogmas superimposed by the intellect and open to the intuition and illumination of the mystics.

This, then, is the ultimate vision of cosmos in Bergson's philosophy. It is a magnificent, all-encompassing vision of humanity, life, and the totality of existence.[7] Even though his articulation of an open society has fallen into

---

[7] It is a vision that inspired Jesuit father Pierre Teilhard de Chardin (1881–1955) to write his widely successful book *The Phenomenon of Man*, published posthumously in 1959. Although he was strongly influenced by Bergson, he was deeply committed to his Christian religion, so his elaboration of pantheism is more restrictive and will not be further pursued here.

oblivion, it is fascinating from our modern vantage point, sixty years later, to see a global village now struggling to embrace all of humanity.

Bergson's earlier writings, especially his text on creative evolution, influenced another great philosopher in the first half of our century: Alfred North Whitehead. Although Bergson (1859–1914) and Whitehead (1861–1947) almost completely overlapped in time, Bergson's major philosophical works predate all of Whitehead's.

Whitehead originally pursued an entire career in mathematics, from 1884 to 1924. Then, at age sixty-three, he accepted an endowed chair at Harvard University, where he taught as a professor of philosophy for thirteen years until his retirement in 1937.

During his years as a mathematician, Whitehead wrote, together with Bertrand Russell, the influential treatise *Principia Mathematica* (1913) His philosophical and metaphysical ideas were expounded much later in his life in his lectures entitled *Science and the Modern World, Religion in the Making,* and *Process and Reality, an Essay on Cosmology.* The last is by far the most difficult of Whitehead's philosophical works, and it brought Whitehead fame as the creator of process philosophy and process theology.

In the preface to *Process and Reality,* Whitehead expresses his indebtedness to Bergson, among others, and in his first chapter (p. 9) he explicitly states that his "philosophy of organism"—his own name for process philosophy—is closely allied to Spinoza's scheme of thought.

However, Whitehead develops and transforms ideas from Bergson and Spinoza (among others) into a thought structure that is fully his own and is marked by his training as a mathematician and scientist. He thus does not emphasize the phenomenon of biological evolution per se, but his process philosophy encompasses both the physicist's world of inert matter and the biologist's world of living matter.

As a scientist Whitehead was deeply impressed by how the Newtonian thought structures collapsed when confronted with the microphysical phenomena of the atom (as described in quantum theory and the theory of relativity).

The advances in physics in Whitehead's generation led to a view of the particles of matter as manifestations of energy fields that suffuse all of space. (We will return to this aspect of matter in our next chapter.) Whitehead thus considers the manifestation of material particles the result of dynamic events in the "invisible" energy field, and he terms these events *processes*.

All events in the material world, be they living or inert, are considered by Whitehead to be a stream of processes, and time is but a "passage of Nature" or a "creative advance."

Much of the inaccessibility of the thought structures in Whitehead's monumental *Process and Reality* lies in his struggle (like Bergson's) to formulate a new conceptual language to grapple with age-old philosophical problems. In the epilogue of one of his last books, *Modes of Thought* (1938), Whitehead examines what he calls "The Fallacy of the Perfect Dictionary." As he puts it: "The critical school of philosophy confines itself to verbal analysis within the limits of the dictionary. The speculative school appeals to direct insight, and endeavors to indicate its meanings by further appeal to situations which promote such insights. *It then enlarges the dictionary*. The divergence between the schools is the quarrel between safety and adventure" (p. 173, italics added). This struggle with the inadequacy of language and the attempt to expand its boundaries is reminiscent of the situation in the fledgling discipline of mechanics in Western civilization in the twelfth and thirteenth centuries. Concepts like force and acceleration were then being forged to enlarge "the Perfect Dictionary."

Whitehead rejects the notion that mind and body are separate entities. In Whitehead's vision they are but separate aspects of the same reality. However, in contrast to Spinoza's static, nonevolving "God and Nature intertwined," Whitehead envisions a *dynamic* pantheistic Universe, where God is creatively participating, is not omnipotent, and yet is the ultimate "ground" with a repository of "eternal objects" in line with Plato's timeless conception (Lowe 1962, p. 42). This repository is what we shall call the "Information Field."

Whitehead's vision of reality implies a Universe that is interconnected but pluralistic; in other words, the Universe is made up of myriad "pulses of experience" that embody material aspects as well as such psychological aspects as "feeling" and "value," concepts that are often totally forbidden in the world of scientific materialism. Perhaps unwittingly, Whitehead affirms Plotinus's idea of the world as one organism, the parts of which are joined by "cosmic sympathy." (See Lowe 1962 on Whitehead's use of sympathy, p. 48.)

Whitehead ultimately incorporates the concept of a "God" in his process theology. Half of this God is primordial in nature, being the repository of eternal forms (that is, the Information Field), and the other half actively participates in creation. As Whitehead states in *Process and Reality:* "Neither God, nor the World, reaches static completion. Both are in the grip of the ultimate metaphysical ground, the creative advance into novelty. Either of them, God and the World, is the instrument of novelty for the other" (p. 411).

And at the very end of his life, less than a month before he died, Whitehead had this to say of "the ultimate quest" (*Dialogues of Alfred North Whitehead*, as recorded by Lucien Price, p. 370):

God is *in* the world, or nowhere, creating continually in us and around us. This creative principle is everywhere in animate and so-called inanimate matter; in the ether, water, earth, human hearts. But this creative principle is a continuing process, and "the process is itself the actuality," since no sooner do you arrive than you start on a fresh journey. Insofar as man partakes in this creative process does he partake of the divine, of God, and that participation is his immortality, reducing the question of whether his individuality survives the death of his body, to the estate of an irrelevancy. His true destiny as co-creator in the Universe is his dignity and grandeur.

Here we find echoes of Spinoza's thoughts in the last two sentences.

The seed idea—God in Nature—originally formulated by Aristotle and later developed by Plotinus and Spinoza, has in our century grown into a more elaborate articulation of pantheism, largely because of Bergson's and Whitehead's familiarity with the results of the natural sciences of their times. This scientific awareness brings something new into pantheism: the joining of natural sciences and metaphysics in understanding the creation of life forms. The work of Bergson and Whitehead implies a search for novel conceptual verbalizations to try to account more profoundly for structural details in the manifestations of the divine creativity in living and nonliving matter—an enlargement of "the Perfect Dictionary."

Considerable developments have taken place in the physical sciences and in microbiology since the advent of Bergson's and Whitehead's philosophies. In Chapters 10 and 11 we will explore further attempts at the harmonization of metaphysics with the physical sciences and with the field of evolutionary biology, and we will draw some conclusions from this.

The belief in a supernatural force operative in Nature, sometimes labeled vitalism or pantheism, has been maligned and ridiculed by most natural scientists in this century. The enormous, majestic rise and success of physics, chemistry, and astronomy in shaping our present worldview—from the innermost part of the atomic nucleus to the farthest reaches of the cosmos—have of course greatly contributed to the rock-hard conviction of most scientists today that "our paradigm knows all."

Curiously enough, such was the situation at the end of the last century, before the works of Planck on the photon, and later in the 1920s and 1930s, with Einstein's theories of relativity and the grandiose unfolding of quantum mechanics. The situation in 1995 is much like it was in 1892. Nuclear physicists have rather neatly wrapped up the four known fundamental forces in Nature into one "super force" that was operative near the beginning of the Big Bang. Apart from a few small problems, everything fits pretty neatly into a coherent conceptual framework for explaining the cosmos of the natural scientist—everything except evolutionary biology! It is here that the great mind-blowing conceptual expansions will take place in the next century.

Furthermore, the expansion of our understanding will likely involve vitalism and pantheism in one form or another and will lead to a synthesis of certain aspects of the humanities and the sciences. We have confined ourselves to looking briefly at Western history, because it is in that arena that modern science has unfolded so spectacularly, and it is there that the thought seeds most germane to the answers we seek may be found.

"The only incomprehensible

thing about the universe is that

it is so comprehensible."

—ALBERT EINSTEIN (1879–1955),
German-Swiss-American physicist

The IDEA of MIND and MODERN PHYSICS

In Chapter 8 we listened to humanistic voices echoing down to us through two millennia. Let us now heed some scientific voices nearer to us in time, through which we may discern that a new worldview is emerging that is fundamentally compatible with the pantheistic seed ideas of the preceding chapter. Whereas the Darwinists remain locked in a material-istic Universe, some physicists are tentatively beginning to embrace the no-tion of a conscious Universe.

This hunger for a new worldview is exemplified in a statement by gifted British author John Fowles in the epilogue of his 1985 novel *A Maggot*:

> I have long concluded that established religions of any kind are in general the supreme example of forms created to meet no longer existing conditions. If I were to ask what the present and future world could best lose or jettison for its own good, I should have no hesitation: all established religion. But its past necessity I do not deny. Least of all do I deny (what novelist could?) that founding stage or moment in all religions, however blind, stale and hidebound

they later become, which saw a superceded skeleton must be destroyed or at least adapted to a new world. (P. 467)

Fowles seeks the pathbreakers for the founding stage of new world models in the "rightlobers." *A Maggot* depicts the duel between the "leftlober,"
the interrogator Ayscough, and the rightlober, the former prostitute Rebecca.
According to Fowles they belong to

> . . . two very different halves of the human spirit. . . . In them-
> selves they are neither good nor evil. Those whom the left lobe (and
> the right hand) dominates are rational, mathematical, ordered, glib
> with words, usually careful and conventional; human society largely
> runs on an even keel, or at least runs, because of them. A sage and
> sober assessor of evolution must regard those dominated by the
> right lobe as far less desirable, except in one or two very peripheral
> things like art and religion, where mysticism and lack of logic are
> given value. . . . They blur, they upset, they disturb. (P. 434)

We recognize here in novelist form Bergson's concept of the intellect (left
lobe) and intuition (right lobe).

In examining the stunning intellectual revolutions in modern physics of
our century, we shall see that new physical models have replaced old ones
and that, paradoxically, in a rational discipline like physics the rightlober
(intuitive) mentality has wreaked havoc throughout the establishment of
classical physicists.

We need only mention a few epochal contributions to demonstrate how
significantly classical physics has been revised. For instance, German physicist
Max Planck in 1900 introduced the concept of the energy "packet" of light, the
photon, to explain the observed characteristics of black-body radiation. We
also note the 1905 paper by Albert Einstein on the theory of special relativity,
which led to the famous $E = mc^2$ equation, as well as the article by Danish scientist Niels Bohr in 1913 that announced the revolutionary new model for the
hydrogen atom. Those three new thought structures severely disturbed the

physical establishments of their times, but eventually, by the force of observa-tions that agreed with the theoretical predictions, they came to be accepted as supplanting and enlarging certain previous aspects of classical physics.

Those aspects concerned the structure and physics of the microworld of atoms and their constituents—the fundamental photon, electron, proton, and neutron particles. Those elementary particles, by and large, are still con-sidered a stable hierarchy, but they are now accompanied by a host of new, short-lived constituents of the atomic nuclei that come and go in the nuclear transformations generated by large atom-smashing accelerators.

Classical physics predicted on the basis of James Clerk Maxwell's elec-tromagnetic theory that an electron (with its negative charge) placed in an orbit around a positively charged nucleus would not stay in the given orbit but would spiral in toward the nucleus. In this spiraling process it would continuously radiate away its energy, much like water flowing out of a bath-tub. Bohr, however, simply postulated, without any proof whatsoever, that the electron would stay put in one of a number of stable orbits around the nucleus. He further postulated that when the atom was disturbed by colli-sions with other atoms or by radiation, the electron would jump from one orbit to another, either emitting or absorbing a given discrete "pellet" of light energy, a photon, in accordance with Planck's hypothesis. With this mathe-matically unproven model, Bohr was then able to explain the discrete col-ors of light emitted or absorbed by agitated hydrogen atoms, a feat the clas-sical physicists had never been able to achieve.

The agreement between Bohr's theory and observations spurred physi-cists to seek theoretical explanations for the existence of the stable electron orbits within the hydrogen atom, and that in turn led to the development of the revolutionary thought structures of quantum mechanics in the 1920s. By this time, through these innovative ideas of Planck, Einstein, and Bohr and subsequent experimental verifications, the photon—the particle of light—was considered one of the elementary particles, like the electron and the atomic nucleus of hydrogen, the proton. The photon traveled through empty space at the velocity of light, had no electric charge, and had no mass when at rest.

Quantum mechanics was the agglomerated result of a series of inspired scientific papers that began with the invention of what is called matrix quantum mechanics by German physicist Werner Heisenberg, who in one inspired day and night created the mathematical theory while on vacation in 1925. That development was followed in 1926 by the invention of wave mechanics in a seminal paper by Austrian physicist Erwin Schrödinger, who developed an alternative mathematical formulation for the revolutionary new discipline of quantum mechanics, or wave mechanics. Six months later German physicist Max Born made a great intuitive leap by postulating that Schrödinger's wave equation describing the electron could be related to an expression that gave the mathematical likelihood of finding the electron at a given point in space and time. Finally, in 1930, British physicist Paul Dirac showed in an elegant mathematical manner by so-called transformation theory that Heisenberg's matrix mechanics and Schrödinger's wave mechanics were two possible mathematical representations for explaining the observational aspects of the physics of the atomic world.

This chapter is not the place to try to convey a comprehensible *physical* meaning of these mathematical theories. In fact, many attempts have been made by brilliant minds, but Nobel Prize–winning physicist Murray Gell-Mann's characterization of quantum mechanics as "that mysterious confusing discipline, which none of us really understands, but which we know how to use" (Bohm and Hiley 1993, p. 1) sums up the futility of that task. In essence, quantum mechanics is an intricate mathematical recipe for predicting very precisely the outcome of a multitude of experimental observations in the microphysical world of atoms and molecules. But nobody truly understands why it works, that is, how it connects *philosophically* to some aspects of reality. This connection—in some manner—comes with the succeeding development of quantum field theories.

The purpose of this chapter is to give an inkling—not in any sense a full understanding—of the revolutionary conceptual (and mathematical) developments in the theoretical physics of our century. The essence of these developments has been to *dematerialize* apparent "solid" matter into spatial and temporal concentrations of energy in underlying invisible field structures that permeate all regions of space. We will also emphasize the "global cho-rus" of researchers involved in these endeavors.

The first three decades of our century, from Planck's introduction of the quantum concept to Dirac's quantum mechanical transformation theory, saw the consolidation of a new worldview of atomic matter. This view ruptured many concepts of classical physics that had built up during the seventeenth and eighteenth centuries. During the 1930s and 1940s quantum mechanical explanations diffused into all branches of the physics of the atomic world, everywhere scoring successes in agreement between new theoretical predictions and more refined observations in the quantum world of the atom.

During that time, however, experimental *nuclear* physics also grew by leaps and bounds. The invention of ever-more powerful atom smashers led to a host of new observational evidence that the world of the atomic nucleus harbored a large number of new elementary particles that required a fundamental expansion of quantum mechanics. That new revolution in the physicists' theoretical thought structures moved in the direction of quantum field theory, an enormously important expansion in philosophical terms of the physicists' conceptual Universe.

The first seeds for the idea that fundamental particles like the photon are manifestations of an underlying electromagnetic field were laid down in several papers. One paper was published in 1928 by German physicists Pascual Jordan and Eugene Wigner, and a series of papers were published between 1928 and 1930 by Werner Heisenberg and Wolfgang Pauli. Those papers described the central dogma of the quantum field theory intellectual approach: "that the essential reality is a set of fields, subject to the rules of special relativity and quantum mechanics; all else (i.e. material particles) is derived as a consequence of the quantum *dynamics* of those fields" (Stephen Weinberg quoted in Pagels's *The Cosmic Code*, p. 269, italics added).

The first of those field theories to be perfected mathematically pertained to one of the four fundamental forces in Nature: the electromagnetic field governing the structure of the atom. Quantum electrodynamics—started by Jordan, Wigner, Heisenberg, and Pauli in the late 1920s—was perfected by American physicists Richard Feynman and Julian Schwinger and Japanese physicist Sin-Itiro Tomonago in the late 1950s and 1960s, for which the latter three were awarded the Nobel Prize in physics in 1969.

In that theory, which successfully amalgamated Einstein's theory of special relativity, quantum mechanics, and electromagnetic field theory, the photon is conceived of as something called a gluon particle through which the electromagnetic force acts and binds the electron in its orbit around the atomic nucleus. The photon scuttles between the electron and the nucleus, keeping it all together in a stable orbit for the electron—thus the name gluon, or "sticking particle," for the photon.

This example again shows how our Newtonian worldview is shattered in the quantum world of the atom. The interior of the atom may *not* be visualized as a miniature "solar system" where the electron revolves in a well-defined orbit around the nucleus under the influence of their mutual electromagnetic attraction.

The second successful field theory concerned the second fundamental force in Nature: the weak nuclear interactions responsible for the radioactive decay of heavy atomic nuclei. That field theory of weak interactions was created by American physicists Sheldon Glashow and Stephen Weinberg and Pakistani physicist Abdus Salam in 1967. In addition, the physicists unified their field theory of weak interactions with that of electromagnetic interactions. They showed mathematically that there would exist another set of gluons, the weak gluons, inside the atomic nucleus, through which the quark fields (the ultimate constituents of nuclear particles) would interact so that the nuclei would decay in a way that matches observations. Those weak gluons, the nuclear analogues of the photons, would not be massless like the photons but would have masses several times that of the proton, the hydrogen nucleus. Three of these weak gluons have been experimentally discovered, although the fourth predicted particle, the Higg's particle, is still at large. For their work on creating the field theory of weak interactions, Glashow, Weinberg, and Salam received the Nobel Prize in 1979.

Quantum chromodynamics, the final field theory to be created, concerns the third fundamental force of Nature (the fourth being gravity), that of strong nuclear interactions, which bind the quarks together to form elementary particles known as hadrons, such as the proton, the neutron, and a host of other, much more short-lived nuclear fragments.

That theory is representative of a group of field theories called gauge field theory. The relativistic version of gauge field theory unifies all three field theories: electromagnetic interaction in the atomic structure, the weak interaction of nuclear radioactive decay, and the strong nuclear interactions involving the quark structure responsible for stable elementary particles such as the proton and the neutron.

The relativistic gauge field theory—also known as Grand Unified Theory (GUT)—arose out of mathematical work on so-called Algebraic Lie Groups carried out by American physicists Chen Ning Yang and Robert Mills in 1954. Those mathematical tools were taken up by Glashow, Weinberg, and Salam and later by Russian physicists Ludwig Fadeev and Vladimir Popov, Dutch physicist Gerhard 'tHooft, American physicist Benjamin Lee, and French physicist Jean Zinn Austin to perfect the beautiful mathematics behind the gauge field theory revolution.

That second formidable conceptual revolution in how physicists think about the structure of the nuclear Universe—and thus the ultimate aspects of inorganic reality—led to a very beautiful picture of a material world in which:

1. The essential material reality is a set of fields.
2. The fields obey the principles of special relativity and quantum theory.
3. The intensity of a field at a point gives the probability of finding its associated quanta—the fundamental particles—as observed by experimentalists.
4. The fields interact and imply interactions of their associated quanta. Those interactions are effected by the quanta themselves.
5. There isn't anything else. (Pagels, *The Cosmic Code,* p. 269)

It is natural, with our human desire for concrete imagery, for us to ask: How are these quantum fields to be visualized? Professor Heinz Pagels has written an eminently readable popularization of the rise of quantum theory in this century entitled *The Cosmic Code: Quantum Physics as the Language of Nature* (1982). Pagels attempts in this book to provide a concrete visual analogy for the quantum fields, in an effort to give some inkling of their characteristics (p. 270).

Imagine a mattress made up of steel springs, all hooked together and extending through all of space known to us. In their hookup these springs form a three-dimensional lattice—gridwork—and let us say that this lattice represents the electron quantum field. When one of these springs is plucked, the spring will vibrate, and this vibration corresponds to a quantum of field energy with which we shall associate the existence of a particle—in this case an electron, because we speak of an electron field. If more springs are plucked into vibration, there will be more electrons manifested, each at the place a given spring has been plucked.

For each type of fundamental particle, such as quarks or photons, there is a separate three-dimensional spring mattress, and these mattresses also run through all of space. Imagine further that the springs of one type of field mattress can be connected to the springs of another type of field mattress by "gluon" springs. Thus, the electron lattice may be linked to the quark lattice and the photon lattice, and we have a set of interacting dynamic quantum fields, for which the observed "particles" are but manifestations of the underlying field energy—spring vibrations at the point in question.

Let this wholly interlaced, "messy" spring structure become invisible with infinitely small springs—then we have arrived at some visual analogy to the quantum field worldview. The possible vibrations of these fields manifest themselves as various particles, which move about in space and interact with each other. As Pagels puts it: "The Universe is a great spawning ground and battlefield of the quanta, according to relativistic quantum field theory" (p. 271).

Pagels also speaks of the atomic world of quantum "weirdness" (p. 64), as contrasted with the well-ordered macroscopic world of Newton, a world of solid objects like rocks and planets. In Newton's world the motions of solid objects follow natural laws, which predict perfectly well where these objects will be in space and time, in the past as well as in the future. These laws have been used in our modern age with superb success, for example, in guiding space probes in their trajectories toward distant planetary targets. The quantum theoreticians of the 1920s and 1930s, however, discovered through observation that the quantum world of the atoms that make up these seemingly solid objects does *not* follow such well-determined and predictable motions at the quantum level.

As Pagels emphasizes, the quantum world cannot be visualized in the way we visualize the large, solid objects in the Newtonian world. There are no "solid" objects at the quantum level. Elementary particles like the electron do not behave like solid objects with a well-determined position in space and time. With their mathematical "recipes" quantum physicists can calculate only the *likelihood* of finding an electron at a given point in space and time. Thus, the well-ordered, highly predictable Newtonian world picture has to be abandoned for the microscopic world of the atom.

Likewise, quantum physicists have had to accept a fundamental limitation of their specific inquiries into what reality is made of: the method of observation influences what physicists observe. Physicists, by their method of observation, encounter a specially slanted "view" of reality.

Thus, a more "humble" attitude has inexorably crept into the mindset of a considerable number of modern theoretical physicists. They have unwittingly reached a philosophical position akin to that expressed by Henri Bergson, as discussed in Chapter 8.

Faced with such a profound limitation concerning scientific inquiry into the nature of reality, the quantum physicist can adopt one of two attitudes. The first is to forget about the philosophical implications of quantum theory and just to use its techniques as an eminently workable mathematical research tool. This is the attitude of the vast majority of quantum physicists today. The second attitude is to look more deeply into the philosophical implications of quantum theory. It is our overriding interest in this book to examine some attempts made in that direction by such physicists as Niels Bohr, Albert Einstein, John Bell, David Bohm, and many others.

Even just the mathematical structure of quantum theory in itself validates the exclamation by British physicist Paul Dirac: "God used beautiful mathematics to create the world" (Pagels 1982, p. 295). And as Pagels puts it in his very illuminating *The Cosmic Code*: "Physicists, irrespective of their beliefs, may invoke God when they feel issues of principle are at stake, because the God of the physicist is cosmic order." If that is true, then the physicists' God bears a strong resemblance to Spinoza's God, wedded into matter in an inextricable web of logical mathematical propositions. Einstein himself is reputed to have called his God "an intelligence of such superiority that compared with it all the systematic thinking and acting of human beings is an utterly insignificant reflection" (Einstein 1935). And when confronted by a rabbi with the question "Do you believe in God?" Einstein replied: "I believe in Spinoza's God, who reveals himself in the orderly harmony of what exists—not in a God who concerns himself with fates and actions of human beings" (quoted in Clark 1971, p. 413).

During the last ten years, at least five philosophical books have been written by professional physicists in which the notion of either God or consciousness is dealt with in detail: *God and the New Physics* by Paul Davies (1983), *The Anthropic Cosmological Principle* by John D. Barrow and Frank J. Tipler (1986), *Conscious Mind in the Physical World* by Euan Squires (1990), *The Conscious Universe* by Menas Kafatos and Robert Nadeau (1990), and *The Mind of God* by Paul Davies (1992).

These well-respected physicists grapple with two fundamental philosophical problems that emerge as a result of recent findings in modern physics. Both problems touch upon the limitations of the scientific method of inquiry into reality. We have seen how Henri Bergson at the beginning of this century criticized the natural sciences in that regard as a professional philosopher, but the criticism is now raised within the physicists' camp. It is, however, only fair to acknowledge that the great majority of physicists do not see any limitations in the scientific method of inquiry. As usually hap-pens in scientific revolutions, a small band of visionaries dares to take the intuitive leap, and the establishment will grudgingly follow when corrobo-rative evidence becomes strong enough.

Astrophysicist John Barrow and physicist Frank Tipler address the prob-lem of mind and matter in *The Anthropic Cosmological Principle,* a rich and variegated discourse on what is known as the Anthropic Principle. In their introductory chapter they point out that the existence of mind constitutes a basic postulate in most philosophical systems, whereas physicists as a rule shun any consideration of mind in their theories.

However, as Barrow and Tipler describe, during the last twenty years a collection of ideas has formed under the name of the Anthropic Cosmologi-cal Principle, which could offer a means of relating mind and mindful observers directly to certain phenomena in the physical sciences. The pro-ponents of that principle argue that the Universe we live in exhibits a re-markable fine-tuning in its deeper structure that allowed life to emerge on our planet and perhaps elsewhere. That fine-tuning can be perceived in many ways and at different levels of intellectual penetration. At the most el-ementary level, the anthropicists point to the remarkable fact that the Uni-verse has survived long enough (that is, it has expanded slowly enough) for life to evolve on our planet. The Universe could just as well have collapsed, say, a billion years after the Big Bang started. In that case there would not have elapsed enough time for galaxies to have formed, let alone stars, and the Universe would have existed with no conscious life forms to observe it.

Furthermore, as we stressed in Chapters 1 and 2, life as we know it is based on only four different elements: hydrogen, carbon, nitrogen, and oxy-gen, with a few other trace elements. We also learned that all those elements except hydrogen have to be synthesized in stellar interiors and then ejected into the interstellar medium when the star reaches the exploding nova stage. For an ordinary star, that explosion happens some 10 billion years after the star is born. Thus, for carbon-based life to exist, the Universe must have ex-panded steadfastly for at least 15 billion years so that life could have evolved undisturbed on planet Earth to produce intelligent observers—us!

In other words, we happen to be born in a Universe that is conducive to the emergence of life.

That is the content of the Weak Anthropic Principle. In Barrow and Tipler's formulation, "The observed values of all physical and cosmological quantities are not equally probable but they take values restricted by the requirement that there exist sites where carbon-based life can evolve and by the requirement that the Universe be old enough for it to have already done so" (1986, p. 16). A corollary of this is that no one should be surprised that we live in such an enormously large Universe, because for us to exist the Universe had to expand for 15 billion years.

Another version, the Strong Anthropic Principle, goes to a deeper level, as embodied in the following statement, also in Barrow and Tipler's formulation: "The Universe must have those properties which allow life to develop within it at some stage in its history" (1986, p. 21). In this formulation those properties of the Universe extend down to the existing values of the fundamental constants of Nature (such as the velocity of light, the gravitational constant, the mass of the proton, and Planck's constant) and to the laws of Nature (the laws of gravitation and relativity and quantum theory).

Paul Davies is one professional physicist who speculates on that principle in his books *God and the New Physics* and *The Mind of God.* Davies gives as an example the case of the strong nuclear force holding together the elementary particles that make up atomic nuclei. If that strong nuclear force were only a few percent stronger, "there would be virtually no hydrogen left over from the Big Bang. No stable stars like the sun could exist, nor liquid water. Although we do not know why the nuclear force has the strength it does, if it did not the Universe would be totally different in form. It is doubtful if life could exist" (*God and the New Physics*, p. 187).

Likewise, as shown by astrophysicist Brandon Carter, a change in the strength of the force of gravitation by as little as one part in 10^{40} would eliminate the physical conditions suitable for the formation of an average star like the Sun. The only stars formed would be blue giants or red dwarfs, ruling out any life that depends on solar-type stars for its existence.

As Davies, Barrow, and Tipler point out, some physicists take refuge in the unproven existence of multiple universes. Our Universe happens to be the one where the numbers and laws come out just right for the appearance of life and observers. But as Davies puts it in *God and the New Physics:*

> If we cannot visit the other universes or experience them directly, their possible existence must remain just as much a matter of faith as belief in God. Perhaps future developments in science will lead to more direct evidence for other universes but until then, the seemingly miraculous concurrence of numerical values that nature has assigned to the fundamental constants must remain the most compelling evidence for an element of cosmic design. (P. 189)

One consequence of the possible existence of a cosmic design is the existence of a designer, or in the theologian's world, a God. Barrow and Tipler in some measure sidestep that issue and prefer to generalize the Strong Anthropic Principle to the Final Anthropic Principle: "Intelligent information processing must come into existence in the Universe, and once it comes into existence, it will never die out" (1986, p. 23).

It is interesting to note, however, that Barrow and Tipler connect the principle to Teilhard de Chardin's concept of the noosphere, the sphere of consciousness. Up until humanity's arrival on planet Earth, the nonsapient life forms expanded to cover Earth in a biosphere. Now humanity—thinking life—in its expansion into a global community, is forming a thought sphere, an intelligence layer covering the planet. According to Teilhard de Chardin's mystical thinking, the layer will coalesce into a supersapient being, the Omega Point, which in his religious interpretation represents the Christ-God.

Barrow and Tipler prefer to take Teilhard de Chardin's concept of the Omega Point as representing the filling of the entire Universe with intelligence. In other words, planetary seeds of intelligent life emerging will coalesce into a supersapient Universe. As they put it: "The basic framework of his [Teilhard de Chardin's] theory is really the only framework wherein the evolving Cosmos of modern science can be combined with an ultimate meaningfulness to reality" (1986, p. 204). And they also assert: "Although the Final Anthropic Principle is a statement of physics and hence ipso facto has no ethical or moral content, it nevertheless is closely connected with moral values. . . . No moral values of any sort can exist in a lifeless cosmos" (1986, p. 23).

We have seen so far that our particular Universe, in its fundamental structure of natural laws, *favors*, or better, *critically conditions* the emergence of intelligent life, of mind—albeit observed to exist by us only on our own planet.

Does the new physics entertain the idea of mind residing as an integral structural element of the Universe itself? That notion forms the background for the title of the book *The Conscious Universe* by Menas Kafatos and Robert Nadeau. The authors of that book focus on the philosophical and existential implications of some extraordinary theorems and experiments in the physics of elementary particles. According to the authors, the physics of the colossally large—the cosmology of the Universe—and that of the remarkably small—the world of the atom—meet in positing the existence of mind as a structural element of the Universe.

As we noted earlier, the development of quantum theory during the 1920s and 1930s represents one of the most brilliant chapters in the intellectual history of this century. The theory has been enormously successful in its applications to the phenomena of the subatomic world. As Paul Davies points out, it has given us the laser, the electron microscope, the transistor, the superconductor, and nuclear power.

What is considerably less known among the public are the far-reaching implications quantum theory has for our philosophical view of reality. Those implications emerged during the 1930s and formed the basis for the famous debates between Albert Einstein, the father of relativity, and Niels Bohr, the father of the quantized atom.

A basic tenet of quantum theory is that the observer's measuring device interferes with the actual physical phenomenon to be observed. That situation is embedded in the famous Heisenberg Uncertainty Principle, which in one application states that the observer can precisely determine either the position of an elementary particle (say, an electron) or its motion but never both characteristics at the same time. According to quantum theory, there exists a certain likelihood that the observer will find the particle at a certain position with a certain motion. The physicist expresses that likelihood by a mathematical expression called a wave function. It is important to note that this uncertainty holds only for a single particle. If you have a large group, like trillions of gas molecules in an enclosure, the group behavior is totally predictable in position and motion.

The classical discussion between Einstein and Bohr referred to that uncertainty problem for a single particle. Einstein's position was exemplified by his famous saying "God does not play dice," by which he meant that the underlying real physical world did not contain any uncertainty in the behavior of the atom. In contrast stood the ghostly quantum world of Bohr, to which most physicists subscribed in the 1930s and also today:

> According to Bohr, the fuzzy and nebulous world of the atom only sharpens into concrete reality when an observation is made. In the absence of an observation, the atom is a ghost. It only materializes when you look for it. And you can decide what to look for. Look for its location and you get an atom at a place. Look for its motion and you get an atom with a speed. But you can't have both. The reality that the observation sharpens into focus cannot be separated from the observer and his choice of measurement strategy. (Davies 1983, p. 103)

This is called the Copenhagen interpretation of quantum mechanics. It asserts that the act of human observation brings into existence in the observer's *mind* the latent property of the object under investigation in the quantum world of the atom. In other words, our human observations carve out and bring into existence only selected properties of the underlying atomic reality—whatever that reality is.

Furthermore, the very act of observation removes the quantum particle from the realm of many possible latent characteristics into the realm of a specific, observed characteristic. This is what is meant by Bohr's *ghost* world of quantum possibilities.

That is a mind-boggling limitation of the power of science to provide understanding. The observer's mind is inextricably linked to the picture science creates of reality. Emerging from modern physics in a much sharpened way is a criticism of the explanatory and exploratory power of science—a criticism we remember Bergson had leveled at the natural sciences at the beginning of this century.

The intimate linkage of the observer to the observed picture of the quantum world has led distinguished physicist John Wheeler to formulate a variant of the Anthropic Principle called the Participatory Anthropic Principle. In Barrow and Tipler's words, "Observers are necessary to bring the Universe into being" (1986, p. 22). What Wheeler means is that human observations are necessary to create an "observer-mode reality" out of the ghost world of quantum possibilities.

What if there existed a mind before people? That question may perhaps be answered as a consequence of some sensational experiments conducted in France in 1982, stimulated by the discussion between Einstein and Bohr in the 1930s about the philosophical implications of quantum theory.

Einstein never could accept the uncertainties implied in the quantum theories. Together with co-workers Boris Podolsky and Nathan Rosen, Einstein in 1935 formulated theoretically the experimental conditions for a test (the EPR experiment) to decide whether Nature followed well-defined rigorous laws in the realm of the atom, laws that implied no uncertainties about the characteristics of the elementary particles involved. According to the three scientists, such elementary particle qualities as spin or momentum were given once and for all independent of any observer.

Bohr, on the other hand, argued that the characteristic of the particle is determined only at the time of observation; the particle does not exist, except as one latent possibility of several characteristics in the undulating invisible quantum field. The act of observational measurement concretizes the particle characteristic.

The EPR experiment ideally would operate with individual light particles—single photons. The experiment was so technically difficult that it was not until the late 1970s and early 1980s that it could be carried out satisfactorily, especially in the work of a French group under the direction of Alain Aspect at the Institute of Optics.

The EPR experiment relied on the known properties of a physical characteristic known as spin. Elementary particles such as photons may spin either clockwise or counterclockwise. If the source of energy from which photons are generated has zero spin, and if two photons are emitted from that source in opposite directions, the laws of physics state that the sum of the photons' spins must be zero—that is, if one photon spins clockwise, the other must spin counterclockwise. For this experiment, it is also important to note the law of physics stating that nothing can travel faster than the speed of light.

The experiments conducted by Aspect's group involved two photons generated by a single emission process for a single atom. The unit of energy out of which the photons came had zero spin.

Here the argument between Einstein and Bohr begins. Einstein stated categorically that each photon was born in the emission process with a given spin and would later be caught in the detection apparatus with this initially given spin.

Bohr contended that the individual photons were not real until they were observed in their individual detection apparatuses. Photon A, in its flight to its detection apparatus, had latent in its quantum field vibration the possibility of spinning either clockwise or counterclockwise. Once the detection apparatus had forced this "ghost" photon to become real by the detection process, the photon had to appear in one definite spin direction, and the other latent spin possibility was thus eliminated for photon A. Suppose photon A shows up with a clockwise spin in one of the detection apparatuses. Photon B must then opt for the counterclockwise spin in order for the sum of the spins to equal zero. That in turn implies that the observational measurement of photon A has influenced photon B *instantaneously*—and not even some sort of signal traveling at light speed could exert influence that quickly.

We could say that the results of the Aspect experiment could just as well be explained if photons A and B *started out* with given opposite spin directions. This was the contention of Einstein, Podolsky, and Rosen.

However, John Bell of the CERN laboratory in Europe had shown in 1964 that—if Einstein, Podolsky, and Rosen were right—verifying the spins at the detector apparatuses would become impossible beyond a certain distance between the detector and the emitting source for the single photons. The photons would be destroyed by colliding with other particles. On the other hand, if the Bohr picture was correct, verifying the spins would still be possible regardless of the path lengths that separated the photon emission source and the detectors.

Bell's predictions were crucially confirmed by the Aspect experiment. The photon spins were verified past the limiting path length. Several other experiments since Aspect's have conclusively shown that Bohr's quantum ghost world interpretation appears to be correct (see article by Chiao, Kwiat, and Steinberg in *Scientific American,* August 1993).

The astounding observational fact emerges that within the underlying quantum field manifestation of physical reality, when one observational measurement touches that reality at a given point in space and time, the effects of that observation can instantaneously—faster than light—touch some aspect of reality elsewhere. The effects on that other aspect can be observed simultaneously with the first event. This mind-blowing fact goes under the name of quantum *nonlocality* effects.

David Bohm, a leading quantum theorist,[1] addresses these issues in his book *Wholeness and the Implicate Order:*

---

[1] It is only fair to point to the ascendancy of David Bohm's theoretical ideas of a deterministic quantum world, wherein the nonlocal quantum effects persist nevertheless. (See article by D. Albert in *Scientific American,* May 1994.)

A centrally relevant change in descriptive order required in the quantum theory is thus the dropping of the notion of analysis of the world into relatively autonomous parts, separately existing but in interaction. Rather the primary emphasis is now on *undivided wholeness,* in which the observing instrument is not separated from what is observed. (P. 134)

In short, the world is not a collection of separate but coupled things; rather it is a network of relations. This remarkable book by Bohm contains a number of references to the ideas of Whitehead (see especially p. 207), and several of Bohm's formulations show the influence of or compatibility with Bergson's ideas (on time, for example, p. 207). But Bohm incorporates and develops these ideas within his own visionary scientific personality, which is also imprinted by scientific ideas of our modern times. His evocative but abstract worldview is partially expressed in this statement: "So we are led to propose further that the more comprehensive, deeper and more inward actuality is neither mind nor body but rather a yet higher dimensional actuality, which is their common ground and which is of a nature beyond' (p. 209).

Another example of the influence of Whitehead's process philosophy on the thinking of the physicist-philosphers of our times can be found in the book *Mind, Matter and Quantum Mechanics* by physicist Henry Stapp of the Lawrence Berkeley Laboratories in California. Drawing upon ideas from Heisenberg, Whitehead, physicist-mathematician John von Neumann, and early twentieth-century psychologist-philosopher William James, Stapp elaborates his own psycho-physical theory of reality, the essence of which is: "The physical world described by the laws of physics is a structure of tendencies in the world of mind" (p. 91). Again we find echoes of Spinoza but in a modern framework where William James's "pure experience" and Whitehead's "process" ideas are incorporated.

One of the most audacious new ventures in the philosophy of physics is undertaken by Menas Kafatos and Robert Nadeau in *The Conscious Universe.* Kafatos is a physicist who specializes in astrophysics, general relativity, and the foundations of quantum theory; Nadeau is a student of cultural history, specializing in the history and philosophy of science. Even though their book is written for the nonphysicist, their writing is on an intellectually demanding level. In contrast to Paul Davies's style of elegant, light popularization, Kafatos and Nadeau first attempt to break new philosophical ground, spurred by the results of the Aspect two-photon experiment. Second, the authors attempt to integrate the radical new ideas of quantum "holism" into today's frontier of philosophic physical thinking.

The cornerstone of their new ideas is the result of the Aspect experiment. The experiment unequivocally demonstrates that wholeness is a fundamental characteristic of nature and that human consciousnesses are inextricable parts of this whole. The daring leap of intuition that Kafatos and Nadeau take is to infer that the entire Universe is in some sense conscious. The authors themselves are highly aware of the intellectual risks that they run by making such a conjecture publicly. In order to meet criticism and even ridicule from their scientific colleagues, they are careful to point out that their idea of a conscious Universe in no way involves an anthropomorphic extension of human consciousness. Rather, they conceive that the cosmic order perceived by human observers is indicative of a "self-reflectively aware" Universe, which through the act of self-reflection reveals the physical order that forms one basis for the manifestation of beings like us. In their opinion this conscious Universe is accessible to us—and only in that sense anthropomorphic— through our basic need to feel a spiritual awareness of unity with the whole.

Kafatos and Nadeau's view of the self-reflectively aware Universe suggests an uncanny resemblance to the fundamental thoughts of Plotinus, whom we discussed in Chapter 8. In Plotinus's philosophical world, Intelligence emanates from the One and is frozen in its flow by a contemplation of the One. The emanation of the Soul from the realm of Intelligence is likewise frozen through contemplation by the Soul of the realm of Intelligence. From the Soul emanates matter, and matter illuminated by the Soul becomes the physical world. Thus, the Soul mediates the imprinting of ideas onto the matrix of indeterminate matter. Perhaps the words of Plotinus were indeed prophetic.

To Kafatos and Nadeau, the scientific realization of the Universe as being a single significant whole may be an indication that humanity has entered a new and more advanced stage in the evolution of consciousness. The authors see the ultimate scientific realization that this Universe is conscious as a new basis for dialogue between science and religion. However, "religious truth like scientific truth must be viewed as a metaphor for that which we cannot fully describe" (p. 181).

From the preceding considerations, it should be evident that we may stand on the threshold of the emergence of a new *scientific* worldview in many aspects more profoundly revolutionary than any other in human history. It may be significant that physicist Paul Davies, in *The Mind of God* (1992), ends his scientific rational discourse with—of all things for a physicist—a consideration of mysticism:

> But in the end a rational explanation for the world in the sense of a closed and complete system of logical truths is almost certainly impossible. We are barred from ultimate knowledge, from ultimate explanation, by the very rules of reasoning which prompt us to seek such an explanation in the first place. If we wish to progress beyond, we have to embrace a different concept of "understanding" from that of rational explanation. Possibly the mystical path is a way to such understanding. I have never had a mystical experience myself, but I keep an open mind about the value of such experiences. Maybe they provide the only route beyond the limits to which science and philosophy can take us, the only possible path to the Ultimate. (F 231)

We can go no further in exploring the frontier philosophical thinking of some highly regarded professional physicists. We have seen that the ideas of consciousness and mind have powerfully come to the foreground as fundamental aspects of reality in the Universe. These ideas are championed by only a minority of physicists, but this minority counts highly respected scientists. Compelling evidence from evolutionary biology may yet add the decisive factor in favor of a conscious Universe.

"In the beginning was the Word,

and the Word was with God,

and the Word was God."

—ST. JOHN THE EVANGELIST
(first century A.D.)

The PLANETARY MIND and INFORMATION

In 1964, C. S. Lewis, the author of *The Screwtape Letters* and the science fiction trilogy *Perelandra* and a professor of Medieval and Renaissance English at Cambridge University, wrote the following passage in his last book *The Discarded Image:*

> The great masters do not take any Model quite so seriously as the rest of us. They know that it is, after all, only a model, possibly replaceable. . . . . It is not impossible that our own Model will die a violent death, ruthlessly smashed by an unprovoked assault of new facts—unprovoked as the nova of 1572. But I think it is more likely to change when, and because, far-reaching changes in the mental temper of our descendants demand that it should. The new Model will not be set up without evidence, but the evidence will turn up when the inner need becomes sufficiently great. It will be true evidence. But Nature gives most of her evidence in answer to the questions we ask her.

The scientific understanding of the Universe as being one singly significant Whole may be an indication that humankind has entered a new and
more advanced state in the evolution of consciousness. C. S. Lewis's prophetic words that the model will change because "far reaching changes in
the mental temper of our descendants demand that it should" may have arrived at their moment of realization. However, this new model has to account for and involve the emergence of life on our planet.

The physicist's God of cosmic-ordered mathematics may serve very well
as far as the inorganic material world is concerned, the world of hadrons and
leptons, the physicist's elementary particles of matter and energy. However,
as noted astrophysicist and cosmologist E. R. Harrison puts it: "A person
who persistently asks where life and mind are in the physical world, will be
passed from science to science like a person with a mysterious illness, who
is passed from specialist to specialist" (*Cosmology*, p. 115).

It seems now, for the first time in human intellectual history, that one
particular speciality can be confronted point blank with the question of life
and mind: the discipline of evolutionary biology. A vast body of information
has accumulated during the past fifty years in that discipline and its associated discipline of microbiology. The time is ripe for a reexamination of explanatory thought structures in light of this new body of evidence.

This will require thinking in what may appear at times radically new directions, but as renegade physicist David Bohm puts it: "Thinking within a
fixed circle of ideas, tends to restrict the questions to a limited field. And if
one's questions stay in a limited field, so also do the answers" (quoted in
C. H. Waddington, *Toward a Theoretical Biology*, 1969).

We have gone to great pains throughout our inquiry in this book to reflect and experience the extraordinary variety of ingeniously contrived
mechanisms of life forms, and we have continually posed the question: Can
blind chance beget such beautifully complex solutions to Nature's biotech-
nological problems? The answer has consistently been no, that there hasn't
been enough time in the available evolutionary timetable. We have also
consistently maintained the absurdity that chance can produce profound
technological knowledge, such as that exemplified in the optics of the eye or
the ultrasophisticated neural network of the human brain.

We have examined the pantheistic thought structures of such great Western thinkers as Aristotle, Plotinus, Spinoza, Bergson, and Whitehead. They all were convinced that there exists a spiritual driving force unfolding the life forms on Earth. Their philosophies exemplify the words of the great French mathematician and mystic Blaise Pascal (1623–1662), a contemporary of Spinoza: "The heart has its reasons of which reason is ignorant." These pantheists are no second-rate luminaries. They all possessed exceptional intellectual powers as well as extraordinarily moral qualities in their life bearings. They were nobles of the soul. Perhaps the time has come to integrate their thinking into the web of modern evolutionary biology.

I believe we are living in the midst of the emergence of a new worldview, what some might call a new religion. Until now, we have held ourselves to be the only intelligent, reflective beings on planet Earth. And, feeling alone, we have spent millions of dollars on projects to try to detect intelligently coded radio messages from outer space. Yet we steadfastly refuse to acknowledge the intelligence displayed in the genetically coded messages right here on Earth. I believe that through all our countless imitations of Nature, we have acted on a deep kinship with a spiritual, conscious, higher intelligence in which we ourselves are embedded.

If chance and natural selection are the only operative mechanisms for evolution, it follows that chance has invented artificial intelligence in human beings. But it is sheer nonsense to assume that that which is not intelligence can beget intelligence. We have used all our intelligence to try to create artificial intelligence in computers. Some believe, like Roger Penrose, that we will never succeed. (See his books *The Emperor's New Mind* and *Shadows of the Mind*.) If our best minds cannot beget intelligence, how can chance do it?

Another unfulfilling aspect of the classical Darwinian theory of blind chance is that it is just as boring, monotonous, and sterile as the steady-state cosmological theory of the Universe. In the Darwinian theory, there is only an endless repetition of shuffle the dice and throw, shuffle the dice and throw. There is no creativity involved—we just wait and complexity increases. It is far more inspiring and challenging to take a leap of intuition and assert that the manifold variables of life forms that we observe in Nature are the result of the creativity of a higher planetary intelligence.

The history of the physical sciences is studded with bold leaps of intuition and imagination. As famous mathematician Henri Poincaré noted in 1913:

> If we were so reasonable, if we were curious without impatience, it is probable that we would never have created Science, and we would always have been content with trivial existence. Thus the mind has imperiously laid claim to this solution, long before it was ripe, even while perceived in only faint glimmers—allowing us to guess a solution, rather than wait for it. (Quoted in Harrison 1981, p. 117)

Albert Einstein, one of Poincaré's contemporaries, stated that "to these elementary laws [of Nature] there leads no logical path, but only intuition supported by being sympathetically in touch with experience." In fact, in a letter to his good friend, philosopher Maurice Levine, Einstein described by a diagram his "postulational method." The scientist should start out with the world of experience and experiment. Then, simply on the basis of physical intuition, the scientist should leap—Einstein called it "the intuitive leap"— to the formulation of an absolute postulate as yet totally unproven. Finally, he or she should proceed to deduce specific theoretical results that can be experimentally verified or discounted (see Schilpp 1951).

Newton made such an intuitive leap when he postulated the existence of the invisible force of gravitation in his famous law of gravitation, which was verified by observations of planetary motions. Bohr made such an intuitive leap when he postulated that the electrons stay in fixed orbits around the proton nucleus of the hydrogen atoms, an idea that was verified by observations of spectral lines. And Einstein made an intuitive leap by asserting that gravity represents a sort of curvature in space and time, a hypothesis that was subsequently verified by the finer details of the motion of the planet Mercury.

In recent years, isolated but forcefully dissident voices have spoken against the validity of Darwinian and neo-Darwinian evolutionary theory (see, for example, Arthur Koestler's *The Ghost in the Machine* and *Janus: The*

Summing Up). Stanley Jaki devotes a chapter of *The Road of Science and the Ways to God* to a comprehensive history of criticisms of Darwinism, including his own.

Eminent astrophysicist Sir Fred Hoyle takes a more straightforward approach in his book *The Intelligent Universe*, wherein on the grounds of mathematical probability theory he totally discounts the Darwinian explanatory scheme and replaces it with a theory that the Earth is continually being bombarded with snippets of genetic code fragments—partial genetic messages—sent out at random in space by a higher intelligence in an unidentified interstellar cloud. Life on Earth is then evolving as those genetic code fragments link up.

Rather than following Hoyle's intuitive leap toward an intelligence in outer space, I prefer to anchor my own intuitive leap to evidence that life has evolved here on Earth during billions of years as the result of ongoing creative activities by an intelligence embedded in our own planet.

The modern Gaia hypothesis represents a somewhat similar intuitive leap. This hypothesis was enunciated by J. E. Lovelock and Lynn Margulis in a paper in the geophysical journal *Tellus* (1973) and was later amplified in Lovelock's 1979 book *Gaia: A New Look at Life on Earth*. Lovelock and Margulis consider the entirety of life forms on Earth as a single entity, one whole, which they name Gaia. They conceive of Gaia as an organic unity, as Lovelock states in his book: "We have assumed that the Gaian world evolves through Darwinian natural selection, its goal being the maintenance of conditions optimal for life in all circumstances including variations in the output from the Sun and from the planet's own interior" (p. 127). However, Lovelock and Margulis do not break with traditional Darwinian dogma, and as such their hypothesis does not resolve the Darwinian probability dilemma.

If eons of time and multitudes of generations are mathematically proven to be insufficient to bring new species into life by the Darwinian hypothesis, is it not fair then to explore an alternative hypothesis? May we not consider the existence of an invisible intelligence field—a Mind Field—that right now exists on Earth along with the physicist's equally invisible vibrating quantum energy fields, which manifest their vibrations as material particles? And that this Planetary Mind Field continually manipulates the material energy field into evolving life forms?

May it not be feasible to explore within an expanding paradigm of modern science the consequences and ramifications of the pantheistic seed ideas we have explored in Chapters 8 and 9? I personally believe so, and I wish in the remainder of this book to share some admittedly personal speculations on these issues. As Alfred North Whitehead emphasized, it is mainly through speculative philosophies that we generally advance our deeper understanding.

One of the first questions to be raised is, how might the Planetary Mind Field manipulate the energy fields of matter to produce biological organisms? I believe we should explore the possibilities in the realm of *light,* that is, in the light field or photon field. In the physicist's world picture, light particles are embedded in the very structure of the atom, acting as intermediaries—gluons—for the electromagnetic force that holds the electron and the nucleus in the atom together.

Throughout human history the phenomenon of light has intrigued scientists and philosophers alike. In the Universe of the modern physicist, light stands out as a unique particle. At rest it has no mass, or weight, and it carries no electric charge, yet it can split itself into an electron-positron pair, and it travels at the ultimate velocity of any particle—the velocity of light. Finally, it is its own antiparticle: light that arises from fields of antimatter is indistinguishable from light given off by ordinary matter.

Physicists have conquered the nature of light mathematically through the discipline of quantum electrodynamics, yet their ultimate understanding of the interaction of light with matter remains nebulous. In quantum electrodynamics, the photon is considered to be a sticking particle that holds the electron and the atomic nucleus together by bouncing back and forth between them. Thus, light is a vital ingredient in all atoms and in the molecules and life forms—including humans—that are made up of atoms.

In Chapter 1 we learned that, according to the physicist's present view of the world, light is by far the most preponderant particle in the Universe. After the Big Bang, which represented the beginning of the history of our Universe, there now resides an afterglow of light so significant that for every one atom of ordinary matter in the Universe there exist 1 billion light parti-cles. Matter is an utterly insignificant "contaminant" in the particle Universe. Although most of the energy resides in matter, almost all particles in the Universe are those of light.

If there is a meaningful scheme to things—and the speed of evolution argues that there must be—then light must somehow be a very important ingredient.

In Chapter 7 we critically examined the Darwinian creation myth and found it to fail at the fundamental level of accounting for how life arises. In Chapter 8 we followed the evolution of the core pantheistic idea, God and Nature intertwined, illuminated by the scientific awareness of various his-torical epochs from ancient Greece to the early twentieth century. And in Chapter 9 we became aware of the revolutionary findings in theoretical physics of our own time, findings that "dematerialized" matter and intro-duced an interconnected, interrelated field web, the nonlocal quality of the quantum energy fields.

In a very deep sense, we carry the imprint of a cosmic interconnected-ness in our physical bodies: All the hydrogen atoms—more than half the matter—in our bodies were once constituents of the Big Bang. The rest of our material bodies (all the chemical elements heavier than hydrogen) were forged in the more "local" crucibles of stellar interiors.

Within our modern scientific framework, it appears feasible to explore the possibility that the light particle—the photon that holds these atoms to-gether and travels endlessly between them—serves a twofold purpose in the Universe. In the physicist's material Universe, it acts as a matter "glue" on the atomic level, but in another sense it may well act as a mediator between the Mind Field and the matter fields. Or perhaps it is the Mind Field. If so, then the photon has a Janus face: Facing the material world it is matter glue, whereas facing the world of intelligence it takes on the characteristics of mind glue or of mind itself.

The idea that light is imprisoned within the human body is a very old one. It was first enunciated by the Persian philosopher Mani, who was born in southern Babylonia about A.D. 215 and was later crucified circa A.D. 276 in Persia. He borrowed extensively from Persian religion and believed that in the beginning there were two principles, one Good and one Evil, God and Satan. Each had its kingdom, the kingdom of Light and the kingdom of Darkness. Satan invaded the kingdom of Light and devoured Primal Man and his five elements, which remained as scattered elements of light in the kingdom of Light. In Adam's body, there was a vast number of those seeds of light, and to Mani, Jesus personifies the kingdom of Light imprisoned in matter. What an extraordinary leap of intuition Mani's thoughts were in view of our modern scientific knowledge concerning light in matter. The amount of light "stored" in the human body is quite astounding. The 100 trillion (10^{14}) atoms in each of our one hundred trillion (10^{14}) cells together store at least 10^{28} photons. This amount of light could illuminate a baseball field for three hours with 1 million watts of floodlights.[1]

Can the science of our times carry us into a deeper understanding of how the Mind Field couples with the light field and thereby manipulates matter to assume life forms? Some exciting new possibilities exist, but these scientific developments represent just a barely perceptible growth point for expanding our thinking in these directions.

In April 1994, at the University of Arizona Health Sciences Center in Tucson, about 300 participants from the disciplines of neurophysiology, quantum mechanics, psychology, and philosophy convened to explore the theme "Toward a Scientific Basis for Consciousness." The philosophical attitudes displayed at this conference ranged from the purely materialistic reductionism (the idea that only atoms and molecules are necessary) of people like Francis Crick and his colleague Christof Koch, to people like mathematician-philosopher David Chalmers with his affinity for William James's radical empiricism and concept of "pure experience." (See *Scientific American,* July 1994, for an excellent review.) Among the many research avenues suggested at this conference, two stand out as particularly relevant to our considerations of a Planetary Mind Field.

---

[1] A revival of Manicheism in Christian thought arose around A.D. 1200. The great English naturalist and bishop Robert Grosseteste (1168–1253) developed a "Metaphysics of Light" in which he intuitively anticipates in embryonic form the concepts of a Big Bang, light pressure, and light as a gluon in matter.

One such avenue is that taken by certain neurophysiologists (including Stuart Hameroff) and certain quantum physicists (including Roger Penrose). Their attention focuses on cell structures called microtubules, which previously were assumed to serve only the function of propping up the cell membrane, that is, to act as a kind of rudimentary cell skeleton. These microtubules are minute tubes of protein, through which electrons or photons may flow. In his book *Shadows of the Mind,* Roger Penrose, an authority on theoretical physics, explores the possibilities that photon fields may reside inside these microtubules and, in their joint interactions, may display nonlocal quantum mechanical effects of the type discussed in Chapter 9. Penrose suggests that these quantum effects are in part responsible for certain aspects of human consciousness. He is very cautious in his conclusions, but he makes no bones about his opinions of the conservative establishment of science: "It is only the arrogance of our present age that leads too many to believe that we now know all the basic principles that can underlie all the subtleties of biological action" (p. 373).

Although Penrose is careful not to involve any concept of mind outside human beings, we may note with great interest his and Hameroff's efforts to work from the biological matter structure via the photon field into the domain of underlying quantum energy fields. This work could possibly result in an eventual opening to the Mind Field.

Another search path of interest explored at this conference was represented by David Chalmers. Whereas Penrose and Hameroff develop a theory of consciousness, so to speak, from the bottom up, Chalmers prefers to go from the top down. He strongly argues that a theory of consciousness must posit the existence of a new fundamental property of reality, namely information. He further asserts that this concept of information has aspects that are both physical and "phenomenal" (or "experiential," that is, subjective). We recognize here a linkage to William James's "pure experience" (see Chapter 11) and Whitehead's "cosmic repository of ideas." In fact, all our own criticism of Darwinian theory, applied at the fundamental level, is based on the question of how Nature accesses Plato's world of ideas.

Whitehead linked his repository of ideas to the creative aspect of a God intertwined with Nature; Chalmers appears to do nothing of the kind. However, his modern approach to information and consciousness could possibly explain how the Mind Field incorporates the world of ideas into our biological life forms.

Another link to the Mind Field could be provided by some of David Bohm's novel ideas in the field of quantum mechanics. His last book (published posthumously) enunciates the idea of *active information* carried by an invisible quantum field that accompanies an elementary particle (Bohm and Hiley, *The Undivided Universe,* pp. 31–38). His basic idea is that a form having very little energy (the quantum field) enters into and directs a much greater energy (the motion of the elementary particle), much as a radio or TV wave containing very little energy can carry considerable amounts of information in the form of sound messages or visual images. Bohm also mentions the biological example of DNA, which carries large amounts of information in its structure (p. 36). Under suitable conditions this form of information precipitates cell growth, but the energy for this cell growth comes from elsewhere in the matter fields.

───────────────

As an experienced natural scientist, I have continually been astounded at the subtle and elegant technological solutions found in the structural details of biological organisms. A fundamental question to be asked is: How does Nature tap into the information needed to construct such biological structures as the radar technology in bats or the infrared telescope "eye" detectors of rattlesnakes? Obviously this information somehow existed millions of years before the appearance of the human mind. The story of evolution is obviously one of an increasing and evolving information content poured into the DNA of the life forms. The technological information embodied in a rose, a rabbit, or a human being was not programmed into the DNA of the first prokaryotic cells. Evolution tells us that there was at first a slow, gradual accumulation of information into

DNA over billions of years, which then erupted into a vastly faster programming of information into the DNA by the creation of multicellular organisms.

Where did this information reside to start with, and how was it accessed? Just as we saw Aristotle's fundamental seed idea, the entelechy, as a precursor to our modern-day concept of DNA, so Plato's concept of a "world of ideas" outside the physical world may be a precursor to the concept of an Information Field residing as a part of reality independent of the physicist's quantum energy fields.

It appears unlikely that the Planetary Mind Field can be identical to this Information Field because, as we have stated before, the creative biological activities of the Mind Field suggest that the field itself evolves, as enhanced, augmented information is poured into emerging life forms.

So far we have not discussed what information really is, although we may have an intuitive feeling for what it represents. The science of information theory is really a very young discipline. It started in earnest with a seminal paper entitled "The Mathematical Theory of Communication," published in 1948 by American physicist Claude Shannon. This paper was later followed by further research in *Science and Information Theory* (1962), written by French-American mathematician and physicist Leon Brillouin. Since then many books and innumerable research papers on the topic have been published, including Jeremy Campbell's *Grammatical Man: Information, Entropy, Language and Life* (1982) and Jeffrey S. Wicken's *Evolution, Thermodynamics and Information* (1987). One of the earliest biologists to apply information theory to biological systems was Lila L. Gatlin in her 1972 book *Information Theory and the Living System*.

A fundamental weakness of the present state of the mathematical development in information theory is that it does not mathematically quantify the *value* of the information communicated to the recipient. In other words, current information theory does not incorporate the subtleties of human language.

However, the discovered deeper underlying relationships between information and energy are clearly relevant to our present discussion. In physics, the second law of thermodynamics states that the ultimate energy state of the Universe is a maximum state of disorder (entropy). *Entropy* is a Greek word meaning "transformation" (see Campbell 1982, p. 37), and the implication of that law is that there will be inexorable flows of heat from higher-temperature regions (say, stars) to surrounding lower-temperature regions (interstellar clouds). Thus, in due time there will be a smooth temperature distribution, popularly called the Heat Death of the Universe. With this final state is associated a maximum state of random (disorderly) motion of all the atoms and molecules in the Universe, which represent a uniform temperature everywhere.

Shannon and later Brillouin were able to show mathematically that information about the state of a physical system of atoms and molecules could be expressed as negative entropy. This idea had been anticipated by Erwin Schrödinger in his searching 1944 inquiry *What Is Life?* In this book Schrödinger makes the point that the atoms and molecules in biological systems are in highly ordered states, whereas the tendency in inanimate matter—especially gases—is to progress toward a state of maximum disorder (entropy).

Shannon and Brillouin showed that information about the state of a physical system can be obtained only if energy is expended to gain such information through observation. Brillouin showed that to gain the yes-or-no information about a molecule being in either of two states required an energy expenditure of approximately 10^{-23} joule/degree Kelvin. By definition this is the energy expended to gain one bit of information (yes or no).

In their article "Energy and Information," Myron Tribus and Edward C. McIrwine ponder the fact that the Sun, in the course of a year, sends out to the Earth in the form of light an amount of energy equal to 3.2×10^{22} joules per degree Kelvin. Translated into bits of information, this energy is equivalent to 10^{38} bits per second.

The verbal information contained in a book of average size is approximately 10^7 bits. According to Carl Sagan (*Cosmos,* p. 270) the total in all the library books of the world is about 10^{17} bits, or 10^{10} volumes. Thus, the energy in the sunshine falling on the Earth each second (10^{38} bits) is equivalent to the information in 10^{30} volumes.

It is important to emphasize that at the present time, we cannot unequivocally assert that light energy contains coded information, but we are seeing tantalizing scientific glimpses of a connection between information and energy.

Former Brookhaven Laboratory scientist Tom Stonier makes the daring suggestion in *Information and the Internal Structure of the Universe* that the photon is made up of two components: an energy component and an information component. Although Stonier applies this idea to strictly physical systems, it is intriguing to view it as a takeoff point for introducing information into biological systems.

What is indisputable is the observed fact that a massive amount of instructional information is embedded in the DNA of biological organisms. We humans are totally used to instructions being written in some kind of human language. However, it is obvious that Nature does not use any *human* language to express the DNA instructions.

In the *New York Herald Tribune* of July 11, 1991, there is an account of the efforts by a group of biologists to apply linguistic methods to look for genetic "word" patterns in the DNA structure—in particular the human genome. These studies, still in their infancy, are trying to decipher "God's own language." Some biologists lean toward the concept of DNA being somewhat like a musical composition. No matter what angle from which we view the problem and in whichever way the genetic instructions are formulated in the DNA, the net outcome is a wondrously complex, information-rich structure such as an orchid, a whooping crane, a dolphin, or a human being. The riddle remains: How is this information placed into the DNA structure?

This Planetary Mind Field proposed here as an answer to that riddle is invisible like its physical counterparts, the gravity and quantum energy fields, and it is distributed throughout the entire planetary structure. In the act of species creation, it focuses its manifestation on certain localized spots on the Earth, be they in the oceans or on land. This higher intelligence has to obey the constraints of the physicist's material world of nuclear particles, atoms, and molecules. It is not omnipotent or omniscient. But it might be able to draw on pools of knowledge that perhaps reside in the energy fields of light.

I f the Planetary Mind Field takes an active role in evolutionary biology, as
I believe, it can do so only by "typing" genetic messages onto the DNA
molecules. Perhaps that typing is done by the photons residing in each
atom, and perhaps—by means at present unknown to us—the atoms can be
correctly placed in the nucleic acids that form the genetic code letters. These
letters can then be arranged into meaningful instructions on the DNA molecules, the genes themselves, and their mutual interaction patterns.

Although the Mind Field may be conceived of as the agent responsible
for inducing genetic variations in life forms, once its message is translated by
the DNA molecule, the resulting life form is on its own in the physical
world, and the Darwinian process of natural selection takes over. The Mind
Field's creativity might be thought of as involving a huge series of creative
trials or sketches that are launched out into an uncontrollable environment.

Almost a century ago, the great Swedish dramatist and amateur naturalist August Strindberg (1849–1912) wrote an essay on the nature of plants
in which the following passage occurs:

> The Creator, this great artist, *who himself evolves during the cre-*
> *ative acts,* in which he makes sketches that he may reject, takes up
> anew unfulfilled ideas, perfects them, and manipulates the primitive
> forms. Truly these forms are created by hand. Often he makes stu-
> pendous advances, by inventing the species, and then science comes
> afterward and "perceives" the existence of gaps or missing links and
> fancies that there have existed in-between forms that now have dis-
> appeared. (Translation of Strindberg 1918; Vol. 47, p. 458; italics
> added)

That conception would, it seems (and as Fred Hoyle has pointed out), eliminate the argument against the existence of a higher intelligence based on
our present impression of an imperfect design. We should *not* expect the
world to be perfectly created by an omnipotent higher intelligence. In fact,
it would be boring if the design turned out to be perfect and static. It is the
quest toward perfection, as we see it in evolutionary biology, that makes for

excitement. The unfolding story of evolution can be viewed alternatively as a frolic, an artistic dance of the superintelligent imagination, or as the pouring of an ever-increasing information content into the ordered structures of living matter.

The trouble with the intuitional leap of postulating the existence of a higher intelligence residing here on Earth is that this particular postulate does not lead immediately to easily verifiable, predicted observational results in the scientist's Universe. If we could witness the creation of a new species we could perhaps verify that the rate of mutations that occurred was accelerated way above chance expectations, and perhaps we could see the genetic message actually being "typed" in by the Mind Field. But alas, that possibility appears remote because of the geological time scales involved.

We may have to live with the uncertainty and unverifiability of our notion of the Mind Field within the present paradigm of natural science. That is not an unusual situation in modern physics. The emergence and success of the probabilistic mathematical tools of quantum mechanics have begun to bring into traditional natural science not only the concept of uncertainty (through Heisenberg's uncertainty relation) but also an element of mystery. Ask quantum mechanists whether they understand the deeper-functioning physics of quantum mechanics and they may well answer with the words of physicist Richard Feynman: "I think it is safe to say that no one understands quantum mechanics. Do not keep saying to yourself if you can possibly avoid it 'But how can it be like that?' because you will go 'down the drain' into a blind alley from which nobody has yet escaped. Nobody knows how it can be like that."

I believe the time has come for science to expand itself to include the idea of intelligence outside ourselves as a vital, active ingredient in the Universe. "Man is the measure of all things" were the words of the Greek philosopher Protagoras (490–421 B.C.). And indeed today these words ring true more than ever. Humans represent the unique life form in which one part has emerged from the realm of inanimate matter but where a second part is emerging conscious of another realm—that of Mind. In our last chapter we will examine how humanity's connection to the Planetary Mind Field leads to the evolution of culture and religion.

"We are all agreed that your

theory is crazy; the question

which divides us, is

whether it is crazy enough."

—NIELS BOHR (1885–1962),
Danish physicist

The PLANETARY MIND
and the EVOLUTION of RELIGIONS

I n *God and the New Physics,* physicist Paul Davies comes to the tentative conclusion that perhaps "science offers a surer path than religion in the search of God." Even though the possible God concept glimpsed in Davies's book is the "physicist's God," the God of mathematically beautiful quantum field theories, Davies has a point in that science, represented by evolutionary biology, may lead us to a "God," that is, to a higher intelligence residing right here on Earth that has playfully acted out evolution.

The further exploration of this idea may require us to embrace a new mode of "reasoning into verity," different from that of the traditional scientists. The great mystic-mathematician Blaise Pascal distinguished between "l'ésprit de la géomètrie" (the mode of reasoning of the natural scientist-rationalist) and "l'ésprit de la finesse" (the mode of reasoning of the humanist-mystic). Bergson distinguished between the intellect and the intuition, and he hoped for and foresaw a day when the findings of the intuition would complement and enrich the findings of the intellect.

It is possible, then, that we are beginning to witness the evolution of our mental faculties into a state that will truly amalgamate intellect and intuition and that we have started down the road to expanding the traditional paradigms within the natural sciences.

We have learned that Bergson hoped the metaphysics of intuition would constructively aid the natural sciences of the intellect. The postulate of a Mind Field being an operative in evolutionary biology might be a first concrete step in that direction.

When the full impact of the idea reaches us that we reside embedded in a conscious planet, that we are embedded *within* a higher intelligence that invisibly surrounds us—like air or gravitation—it will truly affect our every habit of thinking. The religious enthusiast has long been familiar with that idea, but to approach the idea seriously from the rational base of the natural sciences—evolutionary biology—is an altogether different thing. Yet, that vantage point may throw new light on the questions at hand. It enables us for the first time to speculate on the nature of this higher intelligence as it couples to the physical world and to our human world. At the risk of being sunk by rationalist arguments, I "throw myself out into waters 10,000 fathoms deep," as Danish philosopher Sören Kierkegaard has it.

So what else can we deduce of the characteristics of the Mind Field—specifically in its creation of and interaction with the human world? Let me digress briefly into the psychological nature of physicists. The concepts of laughter, joy, and fun are deeply rooted in the nature of many scientists who until now have done the hardest intellectual work within the paradigms of the natural sciences. Pagels, in his book *The Cosmic Code,* writes:

> Physicists take their work seriously. If they did not, probably no one would, since it is so far beyond immediate human experience. But with that seriousness and commitment goes a remarkable sense of play. Without laughter and the joy of creativity the research enterprise would become unbearable. Humor opens the mind. It relieves the tension of concentration and exposes the vulnerability of a merely intellectual comprehension. Physicists love to make jokes about their work and its implications. It occurs to them "the eternal Maker of enigmas" might also be a . . . trickster. (P. 339)

The playfulness of human beings is the subject of a book by Dutch historian and philosopher Johan Huizinga, who in 1935 enunciated this concept in his book *Homo Ludens: A Study of the Play Element in Culture*. His basic thesis is that "play only becomes possible, thinkable and understandable when an influx of mind breaks down the absolute determinism of Cosmos" (p. 3). Huizinga prefers to use the English word *fun* in the book to adequately describe, in his opinion, this concept of playfulness.

I wonder if that sense of fun is not one of the attributes that should be applied to the Mind Field, that in the development of evolution we see an evolving, playing intelligence slowly maturing to the point of creating humankind— a *Deus ludens* (God playing). Humanity is but an imperfect copyist of Nature. Seeing fish, we construct boats. Seeing birds, we construct airplanes. And understanding the energy generation of the Sun, we now try to create fusion. But if those inventions of Nature were not there, we probably would not have tried to invent those things. Might not our essentially playful activities reflect the same qualities in the higher intelligence? I realize that this notion smacks of anthropomorphism—a rather despised term nowadays—yet it may echo a deeper perception: "Man was created in the image of God."

The creative evolutionary work on our planet has taken place over a time span of 3.5 billion years, and the Mind Field's involvement in the creation of humankind has covered, say, 3.5 million years. That implies that the evolution of the human species has been operative for *one-tenth of one percent* of all evolutionary time on Earth. That's a very short time span, relatively speaking, and it should appear obvious that at the moment we are in the focus of creative interest. The Mind Field has been involved in the last branch of the hominids—Homo sapiens—only during the last 100,000 years and with civilized humans only during the last 10,000 or 20,000 years. The inescapable fact is that during 99.9 percent of biological evolutionary time on Earth, God—as symbolized in human religions—was not involved in the affairs of humans, because humans were not yet created. Thus, evolution is far from being a finished chapter, as far as humanity is concerned. On the contrary, *we form a part of an ongoing Mind Field experiment right now, whether or not we choose that role.*

The evolutionary élan vital of the Mind Field is now operative *culturally*. We are trying on various new and ancient political *and* religious life forms, as part of the ongoing experiment.

I have said that this postulated higher intelligence operates under stringent constraints of the laws of the physical world. Unless we are willing to try to think in expanded scientific-realistic terms about such problems as how the Mind Field is linked to humanity, we may never advance significantly in our spiritual evolution.

I see the following constraints on our interaction with the Mind Field. Because of our brains we have, relatively speaking, a free will. I say "relatively speaking" because each individual has a free will, but humanity as a whole does not. As a species, it appears we must evolve according to the directives of the Mind Field—which operates with probability distributions. The statistical average development of humankind is deterministic in the sense that we evolve in the direction the Mind Field tries to go. But the Mind Field is not omnipotent. We may borrow a crude analogy from the physicist's world, involving the piston in an automobile engine. Inside the engine cylinder are combustible gases. According to physical gas laws, every molecule of those gases is free to move in any direction. Some molecules will move upward, some downward, and some sideways. However, when the piston at one end of the cylinder starts to move upward and to compress the gases, the molecules, in addition to their individual free motions, also have to move in one deterministic direction—upward, as determined by the unidirectional motion of the piston.

However, in order for the Mind Field to direct human evolution in *cultural* ways, *it must operate through individual mind channels*. Cultural developments depend on accessing a realm of ideas, of information. Once a culturally influential idea appears in the brain of a single human individual, that individual can act to spread that idea into human society, thus affecting the cultural evolution of the human species.

I believe that the Mind Field can move the individual through its manipulative channel: the subconscious part of the human brain, which is also called the unconscious. Here I am making another intuitive leap, but I am

not alone in doing so. Throughout human history, there has been an awareness of the existence of an unconscious part of the human brain. This par of
the brain has been interpreted in a variety of ways. As Lancelot Law Whyte
puts it in his review article in the *Encyclopedia of Philosophy:*

> The mystics saw it as the link to God; the Christian Platonists as
> a divine creative principle; the romantics as the connection between
> the individual and the universal powers; the early rationalists as a
> factor operating in memory, perception and ideas; the postromantics
> as organic vitality expressed in will, imagination and creation; dis-
> associated Western Man as a realm of violence threatening his sta-
> bility; physical scientists as the expression of physiological processes
> in the brain, which are not yet understood; monistic thinkers as the
> prime mover and source of all order and novelty in thought and ac-
> tion; Freud (in his earlier years) as a melee of inhibited memories
> and desires, the main influence of which is damaging; and Jung as
> a prerational realm of instincts, myths and symbols often making for
> stability. It is natural to seek a common principle underlying these
> partial truths, but we do not possess the unified language in which
> to express it scientifically. (P. 188)

The most penetrating observer and analyzer of human consciousness in
this century is probably Swiss psychiatrist and psychologist Carl Gustav
Jung (1875–1961). He devoted a whole lifetime to such studies, and his
writings were voluminous on matters relating to this subject. His extended
empirical (that is, observational) studies of human consciousness and unconsciousness led him to postulate the existence of a collective unconsciousness, which he believed partly manifested itself in the numerous
myths appearing in various advanced as well as primitive societies throughout human history.

Jung also believed that he had shown through his voluminous studies
that in this collective unconsciousness resided certain fundamental ideas,
patterns of thoughts and images, which he called archetypes. Individuals

receive messages from that inner world through dreams and intuitive flashes. In his later years, Jung became convinced that in this collective unconsciousness resided a God *and that this God needed human collaboration to be fulfilled.* Jung himself experienced this God during many agonizing years of self-exploration.

In light of these considerations, we could postulate that the unconscious in each human being represents a communication channel to the Mind Field, and vice versa.

The view of our being hooked into the Mind Field has some overlap with the views of Lovelock, who in the epilogue of his book *Gaia: A New Look at Life on Earth* muses: "To what effect is our collective intelligence a part of Gaia? Do we as a species constitute a Gaian nervous system and a brain, which can consciously anticipate environmental changes?" (p. 147). My vision of the Mind Field is distinctly different from Lovelock's Gaia. His Gaia is the holistic unity of life on Earth. To me, the Mind Field resides in a mental energy field suffused throughout the entire Earth—the visual manifestation of which may be the life forms created. But those life forms live and die independently of the Mind Field, thrown out into the cruel Darwinian world of natural selection and adaptation.'

Periodically in this book we have encountered the name of William James, especially in conjunction with Bergson and Stapp. In this chapter we shall move him more into the foreground, as he represents a discipline, that of psychology, that has a direct bearing on the mind/matter problem as exemplified in our own human brains.

William James (1842–1910), the brother of author Henry James, grew up in a home where free intellectual growth was strongly encouraged. He obtained his medical degree from Harvard University in 1869. Then followed thirty-four illustrious years of teaching at Harvard University, first in anatomy and physiology in 1873, psychology in 1875, and philosophy from 1879 on. In 1890 he published the two-volume work *Principles of Psychology,* which is still a landmark in the field of empirical psychology. This was followed by *The Varieties of Religious Experience* (Gifford Lectures 1901–1902), *Pragmatism* (Lowell Lectures 1907), and *A Pluralistic Universe* (Hibbert Lectures 1908–1909).

He is still revered in psychological circles for his *Principles of Psychology*. In the *Oxford Companion to the Mind* (Gregory 1987) that book is called "a treasure-house of ideas and finely turned phrases which psychologists continue to plunder with profit. . . . It raises many issues that still challenge scientific enquiry" (pp. 383–396).

Of concern to our discussion here are James's ideas on the religious experience of God. Being a pragmatic psychologist, James valued personal experience as an observational datum, yet he was deeply aware of the pitfalls of trusting that subjective experience to have an objective validity and to represent a connection to some spiritual aspect of an external reality.

This sober attitude toward the experiential facts is vividly demonstrated in *The Varieties of Religious Experience*. That treatise is a broad excursion into the realm of religious experience, with such chapter headings as "Religion and Neurology," "The Reality of the Unseen," "Conversion," "Mysticism," and "Philosophy"—all containing a wealth of case histories of religious experience together with sober analyses of the likelihood that each case attests to a real connection between the individual's mind and "God."

Toward the end of the book, James draws his own conclusion after sifting through all this recorded material. He states an aspiring hope in "the notion that an impartial science of religions might sift out from the midst of their discrepancies (i.e. pantheism, theism, nature and the second birth, works and grace and karma, immortality and reincarnation, rationalism and mysticism) a common body of doctrine which she might also formulate in terms to which physical science need not object" (p. 510).

For James, the anchor point for this embryonic "science of religions" (note James's use of the plural) is the subconscious self in each individual human being. James articulates this belief in the following cautious statement (p. 515): "Disregarding the over-beliefs and confining ourselves to what is common and generic, we have in *the fact that the conscious person is*

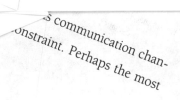

continuous with a wider self through w...

content of religious experi...

tively true as far...

    Jam...

damaging constraint, however, is the existence of evil in the world of human beings. Through the ages evil has been one of the perennial philosophical questions to resolve, and it has been one of the main arguments against the existence of an almighty, all-loving God.[1] We have renounced almightiness in our postulated constraints on the operations of the Mind Field, which does not perform miracles and cannot temporarily suspend the validity of physical laws in the Universe—or even only on Earth. But what about all-lovingness? It should be obvious to everyone who looks around at Nature's creation of life forms that nowhere on Earth does there appear to exist a creative, operative principle of evil. If there were such a malignant cosmic evil, we should see innumerable life forms that actively pursue a path of evil. There is cruelty in the interaction between nonhuman life forms, yes, but the killing is always done to sustain the life of the killer, never for sheer maliciousness and wantonness. Nowhere do we see nonhuman life forms tak... pleasure in torturing to death another life form just for the sake of torturin... The biologist may be able to produce isolated cases, but by and large, am... the millions of living species, the principle of evil cannot be describ... creatively operative. Thus, in the words of Albert Einstein: "Subtle... Lord, but malicious he is not" (Clark 1971, p. 390).

Human beings are the only species that exhibits multifarious... evil. They are the only species that will torture, maim, and mur... sustenance but for sheer evil pleasure. What is it in human bei... duces this behavior? In 1991, L. Morrow published a major... with the title "Evil." Although it gave a broad overview of... and theological aspects of the problem, it failed to mention... pect of evil in human behavior. I believe that we must see... imperfect coupling between the rational part of the h... emotive and reptilian parts.

It is n...
Mind F...
scious in...
the pragm...
produce tan...
take root.
    We can say...
the empirical py...

....
[1] We are concerned here with the "God" of evolutionary biology, not that of the... analogous to Plotinus's the One, whereas the Mind Field may be closer to Plotinu... the presence of a higher intelligence influencing the evolution of life specif... Universe.

That imperfect coupling has been stressed by Arthur Koestler in his book *Janus: The Summing Up*. With the rational parts of their brains, humans can intellectually concoct evil variations of behavior. Some humans then translate these thoughts through their emotive and reptilian parts into evil behavior. The evilness may be triggered by the reptilian part, which generates aggression and the sense of self-survival. If the reptilian and emotive biological parts of the brain sense physical or psychological threats, humans may use the rational parts of their brains to create evil solutions and schemes to reassert their self-esteem during the threatening situation.

Of course, the evil schemes and behavior may also arise from a multitude of other causes, such as the thirst for power and self-aggrandizement or jealousy and passion. But in all this it is the human mind that is the weak link—not the Mind Field. It may be the greatest evolutionary gamble the Mind Field has ever taken to imbed some of its own mind quality into the as yet imperfect and unfinished human brain.

Human suffering is an indisputable concomitant to the existence of evil. It has been a perennially vexing question how to reconcile the existence of human suffering with the existence of an all-powerful and all-loving God. The aforementioned *Time* essay refers to a statement by theologian Frederick Buechner that for three propositions—(1) God is all-powerful, (2) God is all-good, and (3) terrible things happen—if any two of them are true, the third must be false. We have, through our studies of biological evolution, arrived at the conclusion that there exists a Planetary Mind Field that is *not* all-powerful. It is itself evolving in its creativity, and it is restrained to conform to existing physical laws in its biological creativity. Thus, the first proposition has to be abandoned. Maintaining, as we do, the validity of the last two propositions—that "God" is all-loving and that terrible things happen—inevitably requires us to place further restrictions on the capabilities of the Mind Field to intervene directly on individual human suffering. Accepting these two propositions, however, also leads to an increased human responsibility to participate in the evolutionary experiment of which we are part and parcel. *We humans share with the Mind Field a responsibility to alleviate human suffering and gradually extinguish the human element of evil.*

The Planetary Mind Field participates in this endeavor by transmitting through our subconscious individual selves parts of the Information Field. The emergence of moral, religious, and scientific *ideas* into our human world has already dramatically and positively altered our human world. I dare anyone in our privileged Western world to wish to return to the epoch of ordinary human daily lives in the Middle Ages, or for that matter into the world described by Charles Dickens a little over 100 years ago.

German theologian Dietrich Bonhoeffer, who was executed for his participation in an attempt to murder Hitler, came to the conclusion that humankind must act "as if God did not exist." And as we saw earlier in this chapter, Carl Jung believed toward the end of his life that the archetypal God needed the collaboration of human beings for its fulfillment. These are revolutionary religious ideas, but the time may have come when we shall have to face them, even embrace them, and perceive our moral and religious responsibilities in a deeper perspective of cosmological understanding. *We may in fact be at a crossroads in the evolution of religions.*

The idea that humans create evil by their imperfect mastery of the evolutionary gifts of the Mind Field, the rational brain, the emotive brain, and the reptilian brain, one on top of the other, in no way needs to reflect on the attributes of the Mind Field. In contrast, Nature around us bespeaks of its love, and the voices of the mystics—be they from Greece, Western Europe, India, or China—echo through history: "God is Love."

When we peruse the indexes of Davies's *God and the New Physics*, Pagels's *The Cosmic Code*, Hoyle's *The Intelligent Universe,* and Arthur Koestler's *The Ghost in the Machine,* we find no entry for the word *love.* But in our quest of the humanistic past, the concept of love is paramount in the philosophies of Plotinus, Spinoza, Bergson, and Whitehead. It is time that the concept of love be reintroduced into an account of "the real world," together with beauty, truth, and moral values (such as envisioned by Whitehead). Perhaps love is the fifth elementary force in Nature—along with gravitation, electromagnetic interaction, and weak and strong nuclear interactions—and maybe they were all united in the great fireball of the Big

Bang. A similar conclusion is reached at the very end of Amit Goswami's caring book *The Self-Aware Universe* (1993). Goswami, a professor of physics at
the University of Oregon, concludes his book with the words of Teilharc de
Chardin: "Someday after we have mastered the winds, the waves, the tides
and gravity, we shall harness . . . the energy of Love."

We have spoken of the constraints of the Mind Field's interactons
with the real world. What about discerning some temporary goals
for the Planetary Mind Field's creativity in the evolution of humanity? That
may appear preposterous to contemplate, yet I cannot suppress some naive
thoughts on those matters, which I will share at the end of our journey
through this book.

Nobel Prize–winning physicist Stephen Weinberg reflects a strong current of deep human pessimism today when he finds "that the more we know
about the Universe the more it is evident that it is pointless and meaning-
less" (quoted in Pagels 1982, p. 343). However, in Weinberg's opinion. "the
effort to understand the Universe is one of the very few things that lifts
human life a little above the level of farce and gives it some of the grace of
tragedy" (Pagels 1982, p. 312).

I prefer to align myself with the other strong current in human thought,
which is optimistic and is reflected, for example, in the evocative and poetic-
scientific writings of Father Teilhard de Chardin. Like him, I believe we are
actively participating in ongoing evolutionary efforts that, in his view, will
lead to the development of a consciousness sphere, a noosphere, over the
surface of planet Earth. That sphere will ultimately rejoin a Point Omega.
Because of his Christian commitment, Teilhard de Chardin identifies the
Point Omega with Christianity, and to him evolutionary biology is Christ
working through matter. However, I prefer to identify with a higher intelligence that is restricted, maybe even fallible: the Planetary Mind Field. I prefer Bergson's vision of a global community where the goals of an open and

pluralistic society are realized, where the number of mystics grows as evolution progresses in perfecting our communication with the Mind Field, where we have a community based wholly on the principle of love in a new unstructured religion.

Teilhard de Chardin himself admitted that evolution worked in human cultural patterns right now, as well as in the realm of religions. Christianity was the final and most fulfilling religion to Chardin, but it will not be the last phylum on the religious evolutionary tree. Another religion will surely follow that is more global, less ritualistic, and less dogmatic.

If we accept the hypothesis that there exists an invisible Planetary Mind Field in which we are embedded, as we are embedded in an invisible gravitational field and an invisible air medium, certain consequences follow that have a bearing on the issue of religion.

Upon perusing the monumental sixteen volumes (c. 570 pages each) of *The Encyclopedia of Religion* (1987), edited by Mircea Eliade, we find that the Mind Field has connected to individual human brains innumerable times in religiously symbolic ways. This connection has spawned hundreds, if not thousands, of diverse major and minor religions throughout the planet from Paleolithic times to our days. Homo *religiosus* may respond to our intuitive faculty to access the Planetary Mind Field, whereas Homo sapiens accesses the Information Field via the Planetary Mind Field through the intellectual faculty.

Three of the world's major religions today, Christianity, Islam, and Buddhism, claim that only they can provide the sole path to spiritual redemption and access to God, that is, access to the Mind Field. To me, as a scientist, these claims of uniqueness appear absurd in view of the fact that *at least 90 percent of humankind past and present never adhered uniquely to any one of these religions*. Today the Christian religion claims some such figure as 1 billion adherents, or 14 percent of a global population of maybe 7 billion. However, we should bear in mind that according to some demographic estimates, 80 billion humans have existed in the past 4 million years. The Christian religion emerged only 2,000 years ago, and up until, say, A.D. 1500 it was a very local religion within the confines of Europe. Nowhere else on the planet was Christianity known by and large. Thus, for the overwhelming majority of human beings throughout human history, Christianity has not been a path to so-called salvation. The same argument can be brought against any of the other religions.

Consequently, I think every religion has to abandon its claim to uniqueness. We will undoubtedly evolve into a global state of religious pluralism and a general recognition of religious relativism, such that each religion and its sects will have to accept that there are many paths to access the Mind Field. This was very much the belief of William James as he expressed it in *The Varieties of Religious Experience.* "Ought all men to have the same religion? Ought they to approve the same fruits and follow the same leadership? . . . Or are different functions in the organism of humanity allotted to different types of man, so that some may really be better for a religion of consolation and reassurance whilst others are better for terror and reproof? It might conceivably be so and we shall, I think, more and more suspect it as we go on" (p 333).

Have the traditional religions outlived their usefulness, when considered in the context of modern evolutionary biological and cosmological knowledge? According to Robert Wright in his 1992 article "Science, God and Man" in *Time* magazine, there is a significant segment of the population of the Western world that lives in a spiritual limbo, not accepting the traditional faith systems yet not entirely abandoning the belief in some spiritual power and wishing that science would come to their rescue. In this book, I have tried to show that there are a number of scientists who feel forced by their *scientific* convictions to embrace the hypothesis of something akin to a Mind Field. Maybe a new religion will come out of this, or perhaps a transformation—an evolution—of the content in a traditional religion will emerge.

One aspect that indubitably will be stressed more and more as religions evolve is the ecological connection to the human mind. In that sense Lovelock's Gaia concept will need to be incorporated into any new developments. But his concept of the interconnectedness of all life forms on our planet needs to be coupled with the idea of a Mind Field that has created these life forms. Life is a material manifestation of the Mind Field.

One sees the ecological connection emerging within the Christian tradition in the beautifully written book *The Dream of the Earth* (1988) by ecotheologian Father Thomas Berry:

> What I am proposing here is that these prior archetypal forms that guided the course of human affairs are no longer sufficient. Our genetic coding, through the ecological movement and through the bioregional vision, is providing us with a new archetypal world. The universe is revealing itself to us in a special manner just now. Also the planet Earth and the life communities of the earth are speaking to us through the deepest elements of our nature, through our genetic coding.
>
> In relation to the earth we have been autistic for centuries. Only now have we begun to listen with some attention and with a willingness to respond to the Earth's demand that we cease our industrial assault, that we abandon our inner rage against the conditions of our earthly existence, that we renew our human participation in the grand liturgy of the universe. (P. 215)

Karen Armstrong, in *A History of God,* suggests that if any particular idea of God is to survive, it must work for the people who develop it, and that ideas of God change when they cease to be effective. She argues that the concept of a personal God who behaves like a larger version of ourselves was suited to humankind at a certain stage but no longer works for an increasing number of people. Understanding the ever-changing ideas of God in the past, as well as their relevance and usefulness in their time, is a way to begin the search for a new concept for the twenty-first century. She feels that such a development is virtually inevitable in spite of our despair over an increasingly "Godless" world, because it is a natural aspect of humanity to seek a symbol for the ineffable reality that is universally perceived.

Whatever the future evolution of religion holds, it is important to keep in mind that if the hypothesis of the existence of a Mind Field is valid, it strongly implies that this evolution is not a one-sided affair but rather a joint effort between the Mind Field and ourselves. The Mind Field is continually trying to expand our awareness, be it through science, arts, or religion, and to enlarge our consciousnesses so that we emerge as responsible "citizens of the Universe."

My own feeling is that the traditional religions, *by accepting religious relativism,* would meet the religious needs of large segments of humankind for some time to come. Any genuine new religion needs, as Karen Armstrong points out, a true *religious* genius—we have had enough scientific geniuses. But the new religious genius needs to create a new, globally encompassing symbolism of the sacred. Modern secularized society has largely abandoned the notion of sacredness. Unless we can reinstate an awe and a reverence for creation and for its originator, the Planetary Mind Field, and express this through symbolism of the sacred, the new religious efforts will be worthless and full of mere platitudes. Sacred symbolism is needed to shift us out of our everyday existence into a kind of inner sanctum where the connection between the individual self and the Mind Field matters.

The great Swedish novelist and dramatist August Strindberg anticipates some of those ideas in an essay on "The Mystique in World History":

> That the humans do not know what they do is their excuse, but this should also teach them that they are tools in Someone's hands, whose intentions they cannot understand, but who looks to their best. . . . The Great Syntheticist, who unites disparities, solves the contradictions, maintains the equilibrium, is no human being, and cannot be other than the invisible Lawmaker, who in freedom changes the laws according to changing circumstances. The Creator, the Dissolver, the Maintainer, he may then be called by whatever Name. (Translated from Strindberg 1918, Vol. 54, p. 398)

One might say, in looking at human history, that a major goal of the Mind Field has been to create a truly global, religiously pluralistic *community* in the exceedingly short time span (evolutionarily speaking) of 10,000 years and with a growing number of mystics to perfect certain unfinished faculties in humankind.

There may be a vaster goal in the Mind Field, however, as partly anticipated in a memorable essay by biologist-medicinist Lewis Thomas entitled "The Role of Man on Earth":

> Conceivably, and this is the best thought I have of us, we might turn out to be a sort of sense-organ for the whole creature, a set of eyes, even a storage place for thought. Perhaps, if we continue our own embryological development as a species, it will be our privilege to carry seeds of life to other parts of the galaxy.

For 500 million years the major evolutionary thrust on Earth has been upward and outward. First the amphibians began to crawl onto land, then the birds and insects took to the air, and now humans have stepped onto the Moon. In light of humanity's emerging awareness of the possibilities for life elsewhere in the Universe, it would not surprise me if a second powerful goal of the Planetary Mind Field were to establish contact, through the development of human beings, with other Mind Fields in the cosmos. Perhaps the constraints placed on the Mind Field require such a manner of contact.

These two goals of evolution could make human life intensely exciting and vibrant: a global community fostered in love and united in a conquest of space as we are carried on the current wave of evolutionary creativity.

We should look at our individual gift of life as a glorious invitation to participate in the creative cosmic efforts to reach those two goals. It is up to us to accept the invitation, to use the gift of life constructively or destructively. I am convinced that among the 7 billion people on planet Earth, there are many who have unwittingly accepted the invitation, and more will follow.

We should not aspire to forsake this world but rather to live out our roles as women and men have done throughout the ages—to do our jobs honestly and patiently, whatever they are, and to foster good children who will become good human beings. These old-fashioned virtues and ideals are as valid today as yesterday. But now, perhaps for the first time, we may as an emerging global community focus our eyes outward, through science, to join in the search for contact with the cosmic community of intelligences. We may also project our mind-souls inward, through the humanities, in our search for a truer, more genuine contact with our Earth Mother, who with immense patience fashions a Master Weave—us.

In truth we have Dante's vision of "Paradiso" applied to *Earth*. The last line of his final canto ends with the image of "Love which moves the Sun and the other stars."

BIBLIOGRAPHY

Albert, D. Z. 1994. "Bohm's Alternative to Quantum Mechanics." *Scientific American*, May: 58–68.

Alberts, B., D. Bray, J. Lewis, M. Raff, K. Roberts, and J. D. Watson. 1989. *Molecular Biology of the Cell*. New York and London: Garland Publishing Inc.

Armstrong, K. 1993. *A History of God*. New York: Knopf.

Arrhenius, S. 1908. *Worlds in the Making*. New York: Harper and Brothers.

Attenborough, D. 1979. *Life on Earth*. Glasgow: William Collins Sons & Co. Ltd.

Augros, R. M., and G. N. Stanciu. 1984. *The New Story of Science*. Lake Bluff, Ill. Regnery Gateway Inc.

———. 1987. *The New Biology*. Boston and London: Shambala.

Ayala, F. J., and T. Dobzhansky. 1974. *Studies in the Philosophy of Biology*. Berkeley and Los Angeles: University of California Press.

Bacon, F. (1561) 1960. *The New Organon and Related Writings*. (*Novum Organum.*) Indianapolis: Bobbs-Merrill Educational Publishing.

Barrow, J. D., and F. J. Tipler. 1986. *The Anthropic Cosmological Principle*. Oxford and New York: Oxford University Press.

Bergson, H. (1907) 1944. *Creative Evolution*. (*L'Évolution Créatrice.*) New York: Modern Library.

———. 1935. *The Two Sources of Morality and Religion*. New York: Henry Hol & Co.

Berry, T. 1988. *The Dream of the Earth*. San Francisco: Sierra Club Books.

Bohm, D. 1980. *Wholeness and the Implicate Order*. London: Routledge & Kegan.

Bohm, D., and B. J. Hiley. 1993. *The Undivided Universe*. London and New York: Routledge.

Born, M., W. Heisenberg, and P. Jordan. 1926. *Zeitschrift für Physik*, 35: 557.

Brillouin, L. 1962. *Science and Information Theory*. New York: Academic Press

Campbell, J. 1982. *Grammatical Man: Information, Entropy, Language and Life.* New York: Simon and Schuster, Inc.

Chiao, R. Y., P. G. Kwiat, and A. M. Steinberg. l993. "Faster Than Light?" *Scientific American,* August: 52–60.

Clark, R. 1971. *Einstein: The Life and Times.* New York and Cleveland: The World Publishing Company.

Cudmore, L. L. 1977. *The Centre of Life.* New York: Quadrangle New York Times Book Co.

Dalton, J. (1808) 1964. *A New System of Chemical Philosophy.* London: P. Owen.

Darwin, C. (1859) 1993. *On the Origin of Species.* New York: Modern Library.

Davies, P. 1983. *God and the New Physics.* New York: Simon and Schuster, Inc.

———. 1992. *The Mind of God.* New York: Simon and Schuster, Inc.

Dawkins, R. 1987. *The Blind Watchmaker.* New York: W. W. Norton & Co.

———. 1994. "The Eye in a Twinkling." *Nature,* 368: 6901–6902.

Eigen, M. 1992. *Steps Towards Life.* Oxford and New York: Oxford University Press.

Einstein, A. 1905. "Elektrodynamik bewegter Körper." *Annalen der Physik,* 4: 891–921 (initial special theory of relativity).

———. 1917. "Cosmological Considerations of Relativity." *Preussische Akademie der Wissenschaften, Sitzungsberichte.* Part 1, pp. 142–152.

———. 1935. *The World as I See It.* London.

Eliade, M., ed. 1987. *The Encyclopedia of Religion.* 16 volumes. New York: Macmillan Publishing Company.

Fowles, J. 1985. *A Maggot.* New York: New American Library.

Gatlin, L. L. 1972. *Information Theory and the Living System.* New York: Columbia University Press.

Goswami, A. 1993. *The Self-Aware Universe.* New York: G. B. Putnam's Sons.

Goudge, T. A. 1967. "Bergson." *The Encyclopedia of Philosophy,* edited by Paul Edwards, Volume 1: pp. 287–294. New York: Macmillan Company and Free Press.

Gould, S. J. 1985. *The Flamingo's Smile.* New York: W. W. Norton & Co.

———. 1989. *Wonderful Life: The Burgess Shale and the Nature of History.* New York: W. W. Norton & Co.

———. 1992. "Review of *Darwin on Trial.*" *Scientific American,* July: 118–121.

Gregory, R. L. 1987. *Oxford Companion to the Mind.* Oxford and New York: Oxford University Press.

Haken, H. 1988. *Information and Self-Organization*. Berlin, Heidelberg, and New York: Springer-Verlag.

Harrison, E. R. 1981. *Cosmology*. Cambridge and New York: Cambridge University Press.

Hawking, S. 1988. *A Brief History of Time*. New York, Toronto, and London: Bantam Books.

Heisenberg, W. 1925. *Zeitschrift für Physik,* 33: 879.

————. 1974 (in English). *Across the Frontiers*. New York and London: Harper and Row.

Hoyle, F. 1983. *The Intelligent Universe*. London: Michael Joseph Ltd.

Hoyle, F., and C. Wickramasinghe. 1981. *Evolution from Space*. London: Dent.

Huizinga, J. (1938) 1956 (in English). *Homo Ludens: A Study of the Play Element in Culture*. Boston: Beacon Press.

Huxley, J. (ed.). 1954. *Evolution as a Process*. London: Allen & Unwin.

Huygens, C. (1698) 1968. *The Celestial World Discovered, or Conjectures Concerning the Inhabitants, Plants and Productions of the Worlds in the Planets*. London: Cass.

Jaki, S. 1978. *The Road of Science and the Ways to God*. Chicago: University of Chicago Press.

James, W. 1890. *Principles of Psychology*. New York: Henry Holt & Co.

————. 1902. *The Varieties of Religious Experience: A Study in Human Nature*. New York: Longmans, Green and Co.

————. 1907. *Pragmatism: A New Name for Some Old Ways of Thinking; Popular Lectures on Philosophy*. New York: Longmans, Green and Co.

————. 1909. *A Pluralistic Universe: Hibbert Lectures at Manchester College on the Present Situation in Philosophy*. New York: Longmans, Green and Co.

Jastrow, R. 1981. *Enchanted Loom: Mind in the Universe*. New York: Simon and Schuster, Inc.

Johnson, P. E. 1991. *Darwin on Trial*. Washington, D.C.: Regnery Gateway Inc.

Julian of Norwich. (1373) 1966. *Revelations of Divine Love*. Middlesex: Penguin Books.

Kafatos, M., and R. Nadeau. 1990. *The Conscious Universe*. Berlin, Heidelberg, and New York: Springer-Verlag.

Kauffman, S. 1993. *The Origins of Order: Self-Organization and Selection in Evolution*. Oxford and New York: Oxford University Press.

Koestler, A. 1968. *The Ghost in the Machine*. London: Hutchison & Co. Ltd.

————. 1978. *Janus: The Summing Up*. London: Hutchison & Co.Ltd.

Kuhn, T. 1970. *The Structure of Scientific Revolutions*. Chicago: University of Chicago Press.

Lamarck, J. B. (1809) 1984. *Zoological Philosophy. (Philosophie Zoologique.)* Chicago: University of Chicago Press.

Lemaître, G. 1927. "Expansion of the Universe." *Annalles Société, Scientifique de Bruxelles,* 47A:49.

Lewis, C. S. 1964. *The Discarded Image*. Cambridge: Cambridge University Press.

———. 1966. *Till We Have Faces: A Myth Retold*. New York: Time Life Books.

Lewontin, R. 1982. "Adaptation." Chapter 2 in *The Fossil Record and Evolution: Readings from Scientific American (1963–1979)*. San Francisco: W. H. Freeman and Company.

Lovelock, J. E. 1979. *Gaia: A New Look at Life on Earth*. Oxford and New York: Oxford University Press.

Lovelock, J. E., and L. Margulis. 1973. "Atmospheric Homeostasis by and for the Atmosphere: The Gaia Hypothesis," *Tellus,* 26:2

Lowe, V. 1962. *Understanding Whitehead*. Baltimore: The Johns Hopkins University Press.

Lyell, C. 1837. *Principles of Geology*. Philadelphia: J. Kay Jun. & Brother.

MacIntyre, A. 1967. "Spinoza." *The Encyclopedia of Philosophy,* edited by Paul Edwards, Volume 7: pp. 530–541. New York: Macmillan Company and Free Press.

Malthus, T. (1738) 1985. *An Essay on the Principle of Population*. Harmondsworth: Penguin Classics.

Margulis, L., and K. V. Schwartz. 1982. *Five Kingdoms*. San Francisco: W. H. Freeman and Company.

Mayr, E. 1982. "Evolution." Chapter 1 in *The Fossil Record and Evolution: Readings from Scientific American (1963–1979)*. San Francisco: W. H. Freeman and Company.

———. 1982. *The Growth of Biological Thought*. Cambridge, Mass.: Belknap Press.

———. 1988. *Toward a New Philosophy of Biology*. Cambridge, Mass.: Harvard University Press.

——— 1991. *One Long Argument*. Cambridge, Mass.: Harvard University Press.

———. 1994. Interview in *Scientific American,* July: 24.

Merlan, P. 1967. "Plotinus." *The Encyclopedia of Philosophy,* edited by Paul Edwards, Volume 6: pp. 351–359. New York: Macmillan Company and Free Press.

Miller, S. L. 1953. *Science,* 117: 528.

Morrow, L. 1991. "Evil." *Time,* June 10: 38–43.

Newton, Sir I. (1687) 1969. *Mathematical Principles of Natural Philosophy,* translated into English by Andrew Motte in 1729. New York: Greenwood Press.

Nilsson, D., and S. Pelger. 1994. *Proceedings of the Royal Society,* B256: 53–58.

Olson, E. C. 1985. "Intelligent Life in Space." *Astronomy,* July: 10.

Pagels, H. 1983. *The Cosmic Code: Quantum Physics as the Language of Nature.* Toronto and New York: Bantam Books.

Penrose, R. 1989. *The Emperor's New Mind.* Oxford and New York: Oxford University Press.

———. 1994. *Shadows of the Mind.* Oxford and New York: Oxford University Press.

Price, L. 1954. *Dialogues of Alfred North Whitehead.* Boston: Little, Brown.

Prigogine, I., and I. Stenger. 1984. *Order out of Chaos.* New York, Toronto, and London: Bantam Books.

Rebeck, J. 1994. "Synthetic Self-Replicating Molecules." *Scientific American,* July: 48–55.

Russell, B. 1946. *History of Western Philosophy.* London: Allen & Unwin.

Sagan, C. 1980. *Cosmos.* New York: Random House.

Schilpp, P. A. (ed.). 1951. *Albert Einstein: Philosopher-Scientist,* Volume 7. Evanston, Ill.: Library of Living Philosophers.

Schrödinger, E. 1926. *Annalen der Physik,* 79: 361 and 734.

———. *What Is Life?* Cambridge: Cambridge University Press.

Schubert-Soldern, R. 1962. *Mechanism and Vitalism.* Notre Dame, Ind.: University of Notre Dame Press.

Shannon, C. E. 1948. "The Mathematical Theory of Communication." *Bell Systems Technology Journal,* 27:379.

Shapiro, R. 1986. *Origins: A Skeptic's Guide to the Creation of Life on Earth.* New York: Summit Books.

Silverstein, A., and V. Silverstein. 1986. *World of the Brain.* New York: W. Morrow

Simpson, G. G. 1947. *This View of Life.* New York: Harcourt, Brace and World.

Spinoza, B. 1955. *Chief Works.* New York: Dover Publications.

Squires, E. 1990. *Conscious Mind in the Physical World.* Bristol and New York: Adam Hilger.

Stapp, H. 1993. *Mind, Matter and Quantum Mechanics.* Berlin, Heidelberg, and New York: Springer-Verlag.

Stebbins, G. L., and F. J. Ayala. 1985. "The Evolution of Darwinism." *Scientific American,* July: 72.

Stonier, T. 1900. *Information and the Internal Structure of the Universe.* Berlin, Heidelberg, and New York: Springer-Verlag.

Strindberg, A. 1918. *Collected Works.* Stockholm: Albert Bonniers Förlag.

Teilhard de Chardin, P. 1959. *The Phenomenon of Man.* New York: Harper.

Toland, J. 1705. "Socinianism Truly Stated." In *Pantheisticon*. 1720. Reprinted in 1976 by Garland Publishing, New York.

Tribus, M., and E. C. McIrwine. 1971. "Energy and Information." *Scientific American,* September: 179–184.

Tucson Conference on "Toward a Scientific Basis for Consciousness." 1994. Review in *Scientific American,* July: 88–94.

Waddington, C. H. 1969. *Toward a Theoretical Biology.* Edinburgh: Edinburgh University Press.

Wald, G. 1954. "The Origin of Life." *Scientific American,* August.

———. 1994. "The Cosmology of Life and Mind." In *New Metaphysical Foundations of Modern Science,* W. Harman and J. Clark, eds. San Francisco: Institute of Noetic Sciences.

Wallace, A. R. 1903. *Man's Place in the Universe: A Study of the Results of Scientific Research in Relation to the Unity or Plurality of Worlds.* London: Chapman and Hall.

Watson, L. 1980. *Lifetide.* Toronto and New York: Bantam Books.

Weinberg, S. 1977. *The First Three Minutes.* New York: Basic Books.

Whitehead, A. N. 1925. *Science and the Modern World: Lowell Lectures, 1925.* New York: Macmillan Company.

———. 1926. *Religion in the Making: Lowell Lectures, 1926.* New York: Macmillan Company.

———. 1929. *Process and Reality, an Essay in Cosmology.* New York: Macmillan Company.

———. 1938. *Modes of Thought.* New York: Free Press.

Whitehead, A. N., and B. Russell. 1913. *Principia Mathematica.* Cambridge: Cambridge University Press.

Whyte, L. L. 1967. *The Encyclopedia of Philosophy.* New York: The Macmillan Company and the Free Press.

Wicken, J. S. 1987. *Evolution, Thermodynamics and Information.* Oxford and New York: Oxford University Press.

Wright, R. 1992. "Science, God and Man." *Time,* December 28: 38.

INDEX

articulated by physicists like David Bohm and Henry Stapp (see Chapter 9). Tying the vantage point of evolutionary biology into these traditions brings us closer to our notion of the Mind Field.

Because it had to create the unconscious partly out of matter, the Mind Field is constrained by the relative material murkiness of that channel and by the cooperation of the individual in such communication. Thus, the Mind Field has to rely on the statistical properties of those individual communication channels. It must hope that as it sends "waves of ideas" out to billions of individual channels, a very few individuals will be open to receiving the ideas. Humans will then be inspired to make intuitive leaps that fertilize their cultural fields. For example, the mystics occasionally have transparent channels through which total nonverbal communication is possible; the great scientists, on the other hand, have intuitive access to the Mind Field's scientific nature.

In his book *The Cosmic Code,* Pagels relates how famous mathematician Marc Kac distinguishes between two kinds of geniuses, those he calls ordinary and those he calls extraordinary:

> An ordinary genius is someone like you and me except that the genius' ability to concentrate, remember, and create is much greater than ours. Their creative reasoning can be communicated Extraordinary geniuses are quite different. It is not at all clear now they think. They seem to work by a set of rules of their own invention and yet arrive at remarkable insights. They cannot tell you how they got there, their reasoning seems devious. The ordinary genius may have many students. But the devious genius rarely has any, since he cannot communicate his method of solution. (P. 341)

Thus, the murkiness of the individual's unconscious communication channel is the Mind Field's second major operational constraint. Perhaps the most damaging constraint, however, is the *existence of evil in the world of human beings*. Through the ages evil has been one of the perennial philosophical questions to resolve, and it has been one of the main arguments against the existence of an almighty, all-loving God.[1] We have renounced almightiness in our postulated constraints on the operations of the Mind Field, which does not perform miracles and cannot temporarily suspend the validity of physical laws in the Universe—or even only on Earth. But what about all-lovingness? It should be obvious to everyone who looks around at Nature's creation of life forms that nowhere on Earth does there appear to exist a creative, operative principle of evil. If there were such a malignant cosmic evil, we should see innumerable life forms that actively pursue a path of evil. There is cruelty in the interaction between nonhuman life forms, yes, but the killing is always done to sustain the life of the killer, never for sheer maliciousness and wantonness. Nowhere do we see nonhuman life forms take pleasure in torturing to death another life form just for the sake of torturing. The biologist may be able to produce isolated cases, but by and large, among the millions of living species, the principle of evil cannot be described as creatively operative. Thus, in the words of Albert Einstein: "Subtle is the Lord, but malicious he is not" (Clark 1971, p. 390).

Human beings are the only species that exhibits multifarious aspects of evil. They are the only species that will torture, maim, and murder not for sustenance but for sheer evil pleasure. What is it in human beings that produces this behavior? In 1991, L. Morrow published a major essay in *Time* with the title "Evil." Although it gave a broad overview of the sociological and theological aspects of the problem, it failed to mention the *biological* aspect of evil in human behavior. I believe that we must seek the answer in the imperfect coupling between the rational part of the human brain and the emotive and reptilian parts.

....

[1] We are concerned here with the "God" of evolutionary biology, not that of the physicist. The physicist's God may be more analogous to Plotinus's the One, whereas the Mind Field may be closer to Plotinus's vision of the Soul. We are concerned with the presence of a higher intelligence influencing the evolution of life specifically on Earth, rather than a creator of the Universe.